Global Issues Series

General Editor: **Jim Whitman**

This exciting new series encompasses three principal themes: the interaction of human and natural systems; cooperation and conflict; and the enactment of values. The series as a whole places an emphasis on the examination of complex systems and causal relations in political decision-making; problems of knowledge; authority, control and accountability in issues of scale; and the reconciliation of conflicting values and competing claims. Throughout the series the concentration is on an integration of existing disciplines towards the clarification of political possibility as well as impending crises.

Titles include

Berhanykun Andemicael and John Mathiason
ELIMINATING WEAPONS OF MASS DESTRUCTION
Prospects for Effective International Verification

Roy Carr-Hill and John Lintott
CONSUMPTION, JOBS AND THE ENVIRONMENT
A Fourth Way?

John N. Clarke and Geoffrey R. Edwards (*editors*)
GLOBAL GOVERNANCE IN THE TWENTY-FIRST CENTURY

Malcolm Dando
PREVENTING BIOLOGICAL WARFARE
The Failure of American Leadership

Toni Erskine (*editor*)
CAN INSTITUTIONS HAVE RESPONSIBILITIES?
Collective Moral Agency and International Relations

Brendan Gleeson and Nicholas Low (*editors*)
GOVERNING FOR THE ENVIRONMENT
Global Problems, Ethics and Democracy

Roger Jeffery and Bhaskar Vira (*editors*)
CONFLICT AND COOPERATION IN PARTICIPATORY NATURAL RESOURCE MANAGEMENT

Ho-Won Jeong (*editor*)
GLOBAL ENVIRONMENTAL POLICIES
Institutions and Procedures

APPROACHES TO PEACEBUILDING

Alexander Kelle, Kathryn Nixdorff and Malcolm Dando
CONTROLLING BIOCHEMICAL WEAPONS
Adapting Multilateral Arms Control for the 21st Century

W. Andy Knight
A CHANGING UNITED NATIONS
Multilateral Evolution and the Quest for Global Governance

W. Andy Knight (*editor*)
ADAPTING THE UNITED NATIONS TO A POSTMODERN ERA
Lessons Learned

Kelley Lee (*editor*)
HEALTH IMPACTS OF GLOBALIZATION
Towards Global Governance

GLOBALIZATION AND HEALTH
An Introduction

Nicholas Low and Brendan Gleeson (*editors*)
MAKING URBAN TRANSPORT SUSTAINABLE

Catherine Lu
JUST AND UNJUST INTERVENTIONS IN WORLD POLITICS
Public and Private

Graham S. Pearson
THE UNSCOM SAGA
Chemical and Biological Weapons Non-Proliferation

THE SEARCH FOR IRAQ'S WEAPONS OF MASS DESTRUCTION
Inspection, Verification and Non-Proliferation

Andrew T. Price-Smith (*editor*)
PLAGUES AND POLITICS
Infectious Disease and International Policy

Michael Pugh (*editor*)
REGENERATION OF WAR-TORN SOCIETIES

Bhasar Vira and Roger Jeffery (*editors*)
ANALYTICAL ISSUES IN PARTICIPATORY NATURAL RESOURCE MANAGEMENT

Simon M. Whitby
BIOLOGICAL WARFARE AGAINST CROPS

Global Issues Series
Series Standing Order ISBN 0–333–79483–4
(*outside North America only*)

You can receive future titles in this series as they are published by placing a standing order. Please contact your bookseller or, in case of difficulty, write to us at the address below with your name and address, the title of the series and the ISBN quoted above.

Customer Services Department, Macmillan Distribution Ltd, Houndmills, Basingstoke, Hampshire RG 21 6XS, England

Controlling Biochemical Weapons

Adapting Multilateral Arms Control for the 21st Century

Alexander Kelle
Lecturer in International Politics
School of Politics, International Studies and Philosophy
Queen's University Belfast

Kathryn Nixdorff
Professor of Microbiology and Genetics
Darmstadt University of Technology, Germany

Malcolm Dando
Professor of International Security
Department of Peace Studies, University of Bradford, Bradford

palgrave
macmillan

First published in 2006 by
PALGRAVE MACMILLAN
Houndmills, Basingstoke, Hampshire RG21 6XS and
175 Fifth Avenue, New York, N.Y. 10010
Companies and representatives throughout the world.

PALGRAVE MACMILLAN is the global academic imprint of the Palgrave Macmillan division of St. Martin's Press, LLC and of Palgrave Macmillan Ltd.Macmillan® is a registered trademark in the United States, United Kingdom and other countries. Palgrave is a registered trademark in the European Union and other countries.

ISBN-13: 978–1–4039–9372–4 hardback
ISBN-10: 1–4039–9372–6 hardback

This book is printed on paper suitable for recycling and made from fully managed and sustained forest sources.

A catalogue record for this book is available from the British Library.

Library of Congress Cataloging-in-Publication Data

Kelle, Alexander.
Controlling biochemical weapons : adapting multilateral arms control for the 21st century / Alexander Kelle, Kathryn Nixdorff, and Malcolm Dando.
p. cm.
Includes bibliographical references and index.
ISBN 1–4039–9372–6 (cloth)
1. Chemical arms control. 2. Biological arms control. I. Nixdorff, Kathryn. II. Dando, Malcolm. III. Title.

JZ5830.K45 2006
327.1′745—dc22 2006040210

10 9 8 7 6 5 4 3 2 1
15 14 13 12 11 10 09 08 07 06

Printed and bound in Great Britain by
Antony Rowe Ltd, Chippenham and Eastbourne

Contents

List of Tables

Preface

This jointly authored book is concerned with the impact of the revolution in the life sciences on the arms control regimes that have been set up to prohibit chemical and biological weapons. In addressing such a truly inter-disciplinary question we have benefited greatly from exchanges with three different communities of experts: the life-scientists pursuing cutting-edge research in their respective areas of expertise for the betterment of the human condition; the diplomats in The Hague and Geneva involved in implementing the Chemical Weapons Convention and the strengthening of the Biological Weapons Convention; and the numerous colleagues in NGOs and academia with whom we were able to discuss our ideas over the years.

This book is the key outcome of a project on 'Preventive Arms Control: Analysis of the Potential for Arms Control and Verification of Biological Weapons in the Light of New Developments in Biotechnology' (Project PA 600/02), funded by the Deutsche Stiftung Friedensforschung (DSF) and directed by Kathryn Nixdorff. It would not have been possible without the support received from DSF. Malcolm Dando's contribution to this book has been made possible by a research and writing grant from the John D. and Catherine T. MacArthur Foundation on 'Building an Effective Global Prohibition Regime Against Biological Weapons' (grant no. 03–80129–00–GSS). Lastly, Alexander Kelle would like to acknowledge a MacArthur research and writing grant on 'Preventing the Malign Misuse of 21st Century Chemistry – How to Strengthen the Prohibitory Norm Against Chemical Weapons?' (grant no. 05–84295–000–GSS) whose initial phase has informed part of his contribution to this work.

As usual, the opinions expressed are ours alone, as are all remaining errors.

Alexander Kelle/Kathryn Nixdorff/Malcolm Dando
Belfast/Darmstadt/Bradford, January 2006

1
Introduction and Overview

The norm against the deliberate use of poison and disease in warfare can be traced back several hundred if not thousand years. This 'taboo' became embodied in the 20th century in three international treaties which form the basis of the two chemical and biological weapons prohibition regimes that are still today the major instruments in the fight against the spread of biological and chemical weapons proliferation and use. The three legal instruments are the 1925 Geneva Protocol, the 1972 Biological and Toxin Weapons Convention (BWC) and the 1993 Chemical Weapons Convention (CWC).

The 1925 Geneva Protocol came about as a reaction against the misuse of modern chemistry in the form of 'gas' warfare during World War I. It prohibits the use of chemical and biological – or, as in the terminology of the day, 'bacteriological' – weapons in warfare. Not prohibited are for example development and stockpiling of chemical or biological warfare agents. In addition, many states parties to the 1925 Geneva Protocol attached unilateral reservations to their ratifications, which limited the scope of the Protocol even further. During the second half of the 1960s negotiations to comprehensively prohibit chemical and biological weapons (CBW) were separated, which in turn led to the conclusion of the 1972 BWC. While the BWC was hailed as the first multilateral agreement to ban a whole class of weapons of mass destruction, the 1993 CWC has to be regarded as one of the high points of post-Cold War multilateral arms control. The CWC not only bans a category of weapons of mass destruction, but is the first such multilateral treaty that sets up a new international organization for the verification of treaty provisions.

It has become clear over the last few years, however, that the adequacy of the two prohibition regimes which aim at preventing the hostile use of chemistry and biology for offensive military or for terrorist purposes has been seriously called into question. This is due to a series of interrelated events and trends.

(1) The nerve gas attack in the Tokyo subway system in March 1995 has often been called a 'wake-up call,' refocusing attention as to the potential sources of a CBW attack. In this incident, members of the apocalyptical sect Aum Shinrikyo released the nerve agent sarin in several underground trains. The nerve gas attack killed 12 people and injured over a thousand.[1] In addition, the anthrax letters sent through the US mail system shortly after the terrorist attacks in New York and Washington, DC on 11 September 2001 seemed to confirm in a dramatic way that the question of *whether* terrorists can use biological weapons has to be answered in the affirmative and that it is now imperative to think about *when* and *how* such attacks are most likely to occur. Clearly then, terrorist groups had emerged as a new actor in chemical and biological warfare for which the existing control mechanisms were deemed inadequate.

(2) With respect to the BWC, however, the most glaring gap in the controls of this treaty was recognized long before the emergence of the bioterrorist threat. The absence of a verification system that would be able to confirm the treaty compliant behaviour of BWC states parties or, alternatively, uncover violations of the treaty initially triggered the negotiation of so-called Confidence Building Measures (CBM) at the BWC Review Conferences in 1986 and 1991. Also in 1991 the parallel process of strengthening the BWC through a legally binding international instrument, that is treaty or protocol to the BWC, was started with an exercise to first determine the technical feasibility of verification measures for the BWC. This so-called VEREX exercise was followed from 1995 to 2001 by the work of the Ad Hoc Group (AHG) of BWC states parties negotiating what came to be know as the compliance protocol. These negotiations came to an abrupt – and unsuccessful – end in July 2001 when the US government declared the approach taken by the AHG to be the wrong one, an approach to strengthening the BWC which, from the US point of view, would decrease, not increase security.[2]

(3) The chemical weapons prohibition regime is much farther developed than its BW counterpart. Yet, over the course of CWC

implementation a number of problems have come to the fore, the two most important of which relate to first the implementation of several CWC provisions ranging from adherence to CW destruction deadlines to the absence of national implementing legislation. Secondly, there seems to be an unwillingness on the part of a number of CWC states parties to keep the regime up to date with a view to adapting verification provisions to the changing face of the chemical industries worldwide. This does not bode well for agreeing upon and implementing the more far-reaching adaptations of the regime that will be required by the current revolution in the life sciences.

(4) Related to the prohibition of toxic chemicals for weapons purposes is the issue of so-called 'non-lethal' or 'less than lethal' chemical weapons. These chemical incapacitants have long been on the wish-lists of military and security forces. During recent years, however, there seems to have been an increase in interest in toxic incapacitants in the US, Russia and other countries to the point where in this context previously unknown chemical compounds, like the fentanyl-derivative used by Russian security forces to end the theatre hostage-taking in Moscow in October 2002, suddenly appeared on the scene. The development of such calmatives and other incapacitants as offensive agents on one hand creates the illusion of a more humane way of warfare in the future. On the other, the military interest in such chemical compounds threatens to undermine the prohibition of all toxic chemicals for weapons purposes. In addition, the interest in chemical incapacitants is channelling research and development and the build-up of corresponding infrastructures in a direction that might be difficult to distinguish from clearly prohibited activities under the CWC.

(5) Lastly, a series of scientific experiments and their subsequent publication suggests that the range and possibilities for malign use of biology and chemistry have greatly increased. Among the experiments of concern are the unintentional potentiation of poxviruses as a by-product of attempts to develop a mouse contraceptive, and the production of synthetic polio virus from basic chemical compounds.[3]

While these experiments of concern are mostly discussed as yet another variation of the theme of modifying or 'improving' disease-causing agents, there is a different, more fundamental change under way in the life sciences. This paradigm shift is fuelled by the decoding of the human genome and finds its expression in the establishment of new scientific subfields such as systems biology. In practical terms this means that the current scientific and technological revolution in the life sciences changes the focus of the proliferation problem from the chemical or biological warfare agent as the object of malign manipulation to the physiological target in the human body as the object of attack. As the two prohibition regimes that have been set up to address the problem of chemical and biological weapons are agent-based (admittedly in combination with the intended use of these very agents), this revolution in the life sciences cannot but raise the question of the implications this change in our understanding of the human body at the molecular level will have for the normative structure of the two prohibition regimes currently in place.

In general terms the CBW threat is best conceived of as a chemical and biological spectrum ranging from classical lethal chemical warfare agents on one end to toxic industrial chemicals and on to mid-spectrum toxins and bioregulators, and on the other end from traditional to genetically modified biological warfare agents and on through to newly designed agents. It is to be expected that the scope and pace of scientific and technological change in the life sciences will affect all aspects of this spectrum.

There are many unknowns in the future decades. We do not know definitively what actors may use such agents, or when, where and on what scale they may be used and which particular scientific advances should be of most concern. We also do not know what the impact of defensive preparations may be, but we find it a reasonable proposition to assume that as defences improve, those with hostile intent will move from classical/traditional agents to improved (genetically modified) agents and then on to advanced (designed) agents. Should such an offence–defence BW arms race ensue, it is not too difficult to envisage severe scenarios of use: weapons of mass destruction, ethnic weapons, generational attacks and even species threatening agents.

As we will detail in subsequent chapters, the two prohibition regimes in their current shape are ill-equipped to prevent the misuse of scientific and technological advances across the spectrum of the revolution in the life sciences. The goal, then, has to be to adapt the CW and BW prohibition regimes so that they provide an adequate framework for state action and interaction to address the challenges ahead. In this context we understand regime adequacy to be composed of regime effectiveness and regime robustness, two concepts which are interrelated, but not identical. Regime effectiveness on one hand falls broadly within the scholarly debates on international regimes.[4] In these debates there is consensus that the effectiveness of an international regime has two dimensions: first it focuses on the question whether regimes affect state behaviour in the issue area they are set up to regulate. Secondly, regime effectiveness is measured by the impact the regime has on observable data in the issue area. To use an example from the area of environmental politics, it is conceivable that states comply with emission reduction targets set for a particular greenhouse gas – which would satisfy the first aspect of regime effectiveness – and still the regime could have no impact on the ozone layer. In our area of concern the two aspects of regime effectiveness are much more closely related: if states do neither acquire nor use chemical or biological weapons then the goal of prohibiting these weapons has also been achieved – at least if the regime enjoys a universal membership.

On the other hand:

> an international regime has proven *robust* if its members continue to adhere to it and to comply with its injunctions, even after the regime has come under serious stress owing to some outside event that gives some or all of its members a strong incentive to violate, or to use their power to change, central norms and rules of the regime.[5]

To put it somewhat differently, a regime displays robustness when the actors' expectations continue to converge around the regime's normative structure, despite the occurrence of stress factors that challenge the regime. From this, the following five indicators of regime robustness can be extrapolated:[6]

- states continue their membership in the regime;
- successor states of regime members accept the 'inherited' regime membership;
- the majority of states continues to abide by regime norms and rules;
- in order to preserve the integrity of the regime, its members take action against a state which violates regime norms or rules;
- regime members display activities that aim at adapting the regime to the changed environment and thereby secure future adherence to regime norms and rules.

The latter category of activities may well include measures to enhance regime effectiveness. Although effectiveness is not regarded as a function of robustness, there exists a clear causal relationship between the two: a regime which continues to be ineffective in the sense that it does not achieve its proclaimed aims reduces members' incentives to abide by the norms and rules and thus runs the risk of losing its robustness. To strengthen the effectiveness of a regime is therefore an important tool for enhancing regime robustness.

Müller et al. identify a number of stress factors that can undermine a regime's robustness, two of which are of particular importance for our purposes: technological change and shifts in the distribution of power.[7] Many security regimes – including the CBW control regimes – exist in issue areas which are influenced heavily by technological change. Such change can work in two opposing directions: on one hand, technological developments can create new problems which are no longer adequately covered by regime rules and procedures. On the other hand, technological developments might offer new tools for problem-solving, thereby creating the impression that existing instruments have become obsolete. Regardless of its direction, technological change, if left unattended over longer periods of time, can undermine regime robustness and thus necessitate the formulation of new regime norms and rules. The likelihood that such adaptations are made is influenced to a considerable degree by the distribution of power among regime members. Shifts in power distribution can have their origins outside the regime and may well be able to transgress the regime's scope. Or such shifts can be caused by technological breakthroughs in the issue area a regime regulates, which benefits only one or a small group of states participating in the regime.

Our central concern then is with scientific and technological advances in the life sciences that can be expected to undermine the adequacy of the CW and BW prohibition regimes, if these advances are left unattended. In order to address this concern we raise four questions:

1. How are the CW and BW prohibition regimes set up to deal with scientific and technological (S&T) changes affecting the issue areas these regimes are to regulate?
2. What are the areas of concern in terms of S&T advances that might undermine the two regimes' adequacy?
3. How well equipped are the two regimes to deal with the new challenges?

4. Which adaptations of the CW and BW prohibition regimes are needed to bring them into line with the realities of 21st century life sciences?

The first of these questions will be addressed in Chapters 2 and 3, in which an in-depth analysis of the multilateral CW and BW regimes will be provided. Both regimes have their origin in the 1925 *Geneva Protocol for the Prohibition of the Use in War of Asphyxiating, Poisonous or Other Gases, and of Bacteriological Methods of Warfare*. However, in the late 1960s the two regimes were set on different development paths with negotiations for the two categories of weapons being separated and the Biological Weapons Convention (BWC) being successfully concluded in 1972. The CWC, in contrast was concluded only in 1993. In Chapter 2 we will provide a brief discussion of toxic chemicals and the interrelation between developments in chemistry and their misuse for weapons purposes in past CW-programmes. This misuse potential derives from the dual-use characteristics of many toxic chemicals and the equipment used to produce them. Recent changes in the structure of the chemical industry and advances in chemical process technology have reinforced this dual-use aspect. These scientific and technical issues will then be discussed in relation to the implementation of the CWC, where particular emphasis will be placed on the scope and schedules of the CWC, CW disarmament, verifying the permitted use of so-called discrete organic chemicals, transfer controls, and the CWC First Review Conference's performance in addressing S&T related issues. In a similar fashion Chapter 3 will first address biological warfare agents, their characteristics, and the emergence of the biotechnology industries in the latter quarter of the 20th century. The normative framework of the BW prohibition regime, attempts to strengthen the regime in relation to S&T advances since the mid-1970s, and in particular the efforts to negotiate a Compliance Protocol to the BWC will be discussed.

As mentioned above, we are particularly concerned about scientific and technological advances as they relate to the paradigm shift from a focus on various chemical and biological warfare agents as objects of manipulation to the increased understanding of the multitude of ways to interfere with the human body. Hence we use the different control systems in the human body that can be targeted with malicious intent as the ordering principle for our discussion of the scientific and technological advances of concern. In Chapters 4, 5 and 6 we will discuss in detail some particularly important developments in the areas of immunology, neurosciences, and neuroendocrine-immunology that will have a bearing on the evolution of the threat spectrum we will be facing as a result of this paradigm shift.

A number of experiments involving the creation of a killer mouse-pox virus, the transfer of these findings to the cowpox virus, and also the potentiation of a virulence factor of vaccinia virus have made the headlines of scientific publications as experiments of concern that could be misused for nefarious purposes. What these experiments point at, however, is a dramatically increased understanding of the vulnerability of the human immune system. Picking up on this trend of placing greater emphasis on systems biology and the human immune system is the US National Institute of Health's (NIH) prioritization of immunology among the biodefence research activities it is funding. In order to provide some background for the relevance of this increased attention to immunology Chapter 4 will start with an overview of the scientific concepts underlying this field of research. This will be followed by a discussion of immune evasion strategies that microorganisms can pursue. The chapter will lastly outline some possible future threats to the immune system, for whose realization the planned NIH 'encyclopedia' of innate immunity might provide an all too useful blueprint for potential misuse.

Chapter 5 will then analyse advances in neuroscience and asks how these might be misused by somebody with malign intent. Knowledge about the working of the brain and the interaction of various neurotransmitter chemicals with their respective receptors has been accumulated over the last hundred years. What has changed through the genomics revolution, though, is first of all the much more detailed knowledge of receptor sub-type-neurotransmitter interaction, which makes it easier to target specific behaviour, which in turn could be utilized to develop more potent chemical incapacitants than have existed in the past. Secondly, there is an ever greater expectation that not only individual receptor sub-types can be identified and catalogued according to their functions, but that whole control systems responsible for specific behaviours can be elucidated at the molecular level. Thus, again there is a trend toward a systems biology approach discernible. Drawing on examples from the neurobiology of awareness, fear and cognition, but also more automatic homeostatic functions of the human body, the chapter draws comparisons of Cold War research based on the knowledge of the 1950s and 60s with what the biotechnology revolution will enable actors with malign intent to accomplish today or in the foreseeable future in the development of biochemical agents that produce controllable effects on behaviour.

While Chapters 4 and 5 analyse developments in immunology and the neurosciences individually, Chapter 6 will add another level of complexity and bring together these two areas and the new knowledge currently being created in cutting edge research and development. Hence we will look into both the neural regulation of the immune system, including possibilities for malign manipulation, and the immune regulation of the nervous system. The chapter closes with a note of caution that the 'traditional' way of approaching the nervous system, the immune system, but also the endocrine system, as separate entities might not be the most conducive approach in order to come to terms with the misuse potential of any of these systems and the spill-over among them.

Chapter 7 will summarize the current revolution in the life sciences, thus broadening again our discussion of the selected areas presented in the preceding three chapters. This will be contrasted with the evolution of the two prohibition regimes which takes place in slow motion. This will enable us to draw together past performance of the BW and CW prohibition regimes and assess their adequacy in light of the S&T advances discussed in the preceding three chapters.

Finally, in Chapter 8, we put forward suggestions as to how the control system should be developed to match the coming threat to the CW and BW prohibition regimes. This discussion covers both national and international measures to be adopted. In the context of the latter we propose the establishment of a Framework Convention for Biochemical Controls (FCBC). Such a Convention becomes necessary as other measures either being implemented or proposed on the international level do not take into account the paradigm shift with which we are concerned. It should be understood that we accept that all of life's physiological processes will become open to (benign or malign) manipulation as the century progresses, that the threat of massive malign misuse of the biotechnology revolution is severe, and that what we are proposing is an agenda of control system development and implementation that will take some decades to perfect.

2
Science, Technology and the CW Prohibition Regime

1. Introduction

This chapter will analyse the chemical weapons (CW) prohibition regime with a view to the impact that technological characteristics of and developments related to toxic chemicals as well as developments concerning chemical processes have on the control efforts by states parties to the regime. The analysis starts from the hypothesis that recent developments in modern biotechnology, especially the utilization of combinatorial chemistry in for example the pharmaceutical industries of developed countries pose a risk to the international regime set up for prohibiting chemical warfare agents. In order to prevent the CW prohibition regime from being undermined by these – and other – recent developments, a rethinking is needed of the interrelation between the scientific and technological basis of the issue area and the political-legal regime structure brought in place to control the dangers emanating from known chemical warfare agents and other toxic chemicals and biochemicals that could be misused for warfare or terrorist attacks.

The chapter is divided into two substantive parts, the first of which will begin with a discussion of toxic chemicals and some of their characteristics that have made them attractive as chemical warfare agents. As the relationship between scientific and technological progress in chemistry and the military application of toxic chemicals has been a close one for some time, the second section of the first part will provide an outline of the development of chemistry and chemical technology on one side and its misuse in past CW-programmes during the 20th century on the other. The third section will highlight technical issues in CW destruction, while the fourth section discusses dual-use aspects of toxic chemicals as they relate to the verifiability of the peaceful applications of toxic chemicals and the transboundary transfer of such chemicals. The first part concludes with an overview of trends in chemical industry at the turn of the century.

The second part will analyse the interrelation between scientific and technical issues on one side and the negotiations and the implementation of the Chemical Weapons Convention (CWC) on the other. To this end it will be subdivided in five sections dealing with (1) the scope and the schedules of the CWC, (2) chemical weapons disarmament, (3) the verification of the permitted uses of toxic chemicals, including unscheduled chemicals containing phosphor, sulphur or fluorine, also called discrete organic chemicals (DOCs), (4) controlling the transfer of scheduled chemicals to both state and non-state parties to the CWC and (5) the CWC Review Conference's review of scientific and technological development. This latter section will take as its point of departure the CWC's Article VIII, paragraph 22, which gives the First Review Conference a clear mandate to consider scientific and technological issues. It states that:

> [t]he Conference shall no later than one year after the expiry of the fifth and the tenth year after entry into force of this Convention, and at such other times within that time as may be decided upon, convene in special sessions to undertake reviews of the operation of this Convention. *Such reviews shall take into account any relevant scientific and technological developments.* At intervals of five years thereafter, unless otherwise decided upon, further sessions of the Conference shall be convened with the same objective. [emphasis added]

In relation to the preparation and conduct of the Review Conference, both activities of states parties and the OPCW as well as contributions from non-governmental organizations (NGOs) will be considered. The chapter will conclude with a summary of the argument, and will point out the inadequacy of the current regime structures for preventing the malign misuse of 21st century chemistry.

2. Toxic chemicals, chemical technology and military uses of chemicals for weapons purposes

2.1. Toxic chemical plus malign intent equals chemical warfare agent

The term 'chemical warfare agents' ideally would comprise all toxic chemicals that have been developed, produced, or used in a military context with the intention of utilizing its toxicity to man, animals or plants as its primary weapons characteristic. This definition excludes a considerable number of toxic chemicals used in a military environment, which serve other purposes: a case in point is rocket fuel, which is highly toxic, but whose primary purpose is the propulsion of a missile. On the other hand, this definition goes beyond that used in the Chemical Weapons Convention (CWC): according to the CWC's Article II, para. 2 only those chemicals 'which through its chemical action on life processes can cause death, temporary incapacitation or permanent harm *to humans or animals*' (emphasis added)[1] count as chemical warfare agents.

Yet, the CWC's description points to an important functional distinction of chemical warfare agents. They can be used with the intention to kill, harm permanently or incapacitate temporarily. At the same time, there is no clear-cut distinction between lethal and so-called non-lethal chemical warfare agents. Rather, 'there is a gradual increase in the probability of causing death, as the dose increases'.[2] The probability of a CW agent being lethal or non-lethal, in turn, depends on the toxicity of the agent, its mode of employment, and the target's susceptibility/responsiveness to the agent.[3]

Toxic chemicals that have been developed, produced and used as CW agents are usually subdivided into four categories: pulmonary toxicants, blood agents, vesicants or blister agents, and nerve agents.

Pulmonary toxicants: Pulmonary toxicants, sometimes referred to as lung irritants or choking gases, such as chlorine (Cl) or phosgene ($COCl_2$) were the most widely used CW agents during World War I. When inhaled, phosgene in lower doses causes a transitory irritation of the mucous membranes of the respiratory tract. With a delay of between 1 and 24 hours after exposure patients develop 'bronchiolar constriction, acute pulmonary inflamation, pulmonary edema'.[4] In addition, necrosis of bronchial and lung tissue develops. Through the destruction of lung tissue, increasing amounts of blood plasma gather in the lungs, with their capability to provide for oxygen exchange decreasing simultaneously. Death eventually occurs through suffocation.

Blood agents: Blood agents like hydrogen cyanide (HCN) or cyanogen chloride (ClCN) were first used as chemical warfare agents in World War I. However, their high volatility made it impossible to produce them in high enough concentrations on the open battlefield, which led to their replacement by other agents. Blood agents derive their name from their interaction with enzymes responsible for oxygen uptake from the blood, or the transfer of carbon dioxide back from tissue cells to the blood. Symptoms vary according to route of poisoning and dose level. High doses of respiratory intake of hydrogen cyanide can lead to sudden unconsciousness and subsequent respiratory failure leading to death. 'Lower concentrations may produce tachypnea, restlessness, headache, and palpitations followed by seizures, coma, and death.'[5]

Vesicants: Two categories of vesicants or blistering agents have to be distinguished: one are the mustard agents, the other are a group of arsenic agents, like the so-called Lewisite. Both were extensively used during World War I, and in the case of mustard gas is still considered a major CW agent. Blistering agents are almost colourless and odourless, so that detection by the human senses is almost impossible before the onset of symptoms. Yet, depending on the route of exposure and the concentration of the agent, there can be a time lag of between 1 and 24 hours before symptoms appear. During that time tissue damage – either of the skin, mucous membranes, or the lungs – can have progressed to an extent that either long-term hospitalization is required or the victims will die from their injuries. Up to now, no specific therapy to treat mustard casualties exists.[6]

Nerve agents: As their name implies, nerve agents attack the nervous system of the human body, not other tissues. This group of agents shows by far the highest level of toxicity. Exposure can occur through the inhalation of nerve agent vapour or dermal exposure to the liquid form of the agent. These organophosphorous compounds – like tabun, sarin, soman and VX – mainly act by blocking certain neurotransmitters. The effects of exposure to nerve agents can range from nausea and vomiting, to muscular seizures, and severe damage to the central nervous system, resulting ultimately in death. The onset of symptoms can take anything from seconds to a few minutes – in the case of nausea and vomiting after inhalational exposure – to one or more days – in the case of low-level dermal exposures producing effects in the nervous system.[7]

2.2. Chemistry and chemical warfare during the 20th century

2.2.1. Pre-World War I chemistry and CW use during the war

The use of poisons in warfare is recorded all through history in various cultural contexts.[8] However, even by the late 19th century, when more and more chemicals were produced in quantity, including toxic chemicals, their military utilization was not immediately directed towards an exploitation of their toxicity for weapons purposes. Instead it was first:

> directed towards producing better military explosives, as the nitration of natural substances ... was followed by the synthesis of nitro compounds (nitroglycerine, nitrotoluene), but not towards poisonous gases for war.[9]

Thus, it was only with the beginning of World War I when another aspect of the industrial revolution in chemistry had this effect.[10] As Robinson points out, the 'technology initially responsible' for bringing 'toxic weapons out from their prehistory' was the 'large-scale liquefaction of chlorine gas and its packaging into pressure cylinders'.[11]

It does therefore not come as a surprise that when large-scale use of chemical weapons occurred first, that it was chlorine which was used: almost 150 tons of which were released by the German army on 22 April 1915 near Ypres on the Western front. By late summer 1915 the British and by early 1916 the French forces were able to use chlorine gas in the same fashion, that is through the release from gas cylinders, where the gas was carried by the wind to the enemy troops.[12] Subsequently chlorine was replaced by more toxic chemical warfare agents, the first of which was phosgene. Yet another, from a military perspective more effective, chemical warfare agent was introduced in 1917 with the less volatile mustard 'gas', which accounted for most of the casualties due to CW use during the war. Strictly speaking, because of its lower level of volatility, sulphur mustard is not a gas, but has the military advantage of affecting the targets not only via the inhalational route, but also through the skin. Besides the

Table 2.1: CW production and use during World War I

Country	**agent tons produced**	**agent tons used**
Germany	68,100	57,600
France	36,955	28,859
United Kingdom	25,735	15,700
USA	6,215	1,100
Russia	3,650	5,200
Austria	5,245	8,800
Italy	4,100	6,350
Total	**150,000**	**124,200**

Source: Martinetz, *Vom Giftpfeil zum Chemiewaffenverbot*, p. 103.

increases in toxicity of the warfare agents developed and used, their mode of deployment changed. The original gas clouds, which were dependent on the wind blowing in the right direction, were replaced by artillery shells, which could be delivered behind enemy lines. By the end of World War I practically all belligerents had engaged in CW use. In sum, some 150,000 tons of chemical warfare agents were produced, more than 80 per cent of which were consumed on the battlefields (Table 2.1).

The move from chlorine to phosgene and sulphur mustard as chemical warfare agents of choice during World War I was at least partly prompted by the concomitant development of defensive measures against CW. Effects of the highly volatile chlorine and phosgene weapons could be reduced considerably by equipping troops with respirators or gas masks. However, with the increasing use of less volatile and more persistent agents such as sulphur mustard, which equally acts through the skin, the need for protective clothing arose. Yet, the technical problems involved were considerable and as one observer concludes 'by 1918 there was no accepted general issue of anti-gas clothing'.[13]

2.2.2. The interwar years and World War II

The widespread CW use during the war prompted efforts to control the availability of this type of armament. Most notably, in 1925 the *Conference for the Supervision of the International Trade in Arms and Ammunition and in Implements of War* negotiated the *Protocol for the Prohibition of the Use in War of Asphyxiating, Poisonous or Other Gases, and of Bacteriological Methods of Warfare*. This so-called 1925 Geneva Protocol represents a first, albeit limited step, in the direction of regime-building in the issue areas of chemical and biological weapons.[14] The major deficiencies of the Protocol lie in the fact that it applies only in times of war, that it does prohibit only the use of CW and not development, production or stockpiling, and that many of the instruments of ratification by states parties were accompanied by reservations, making the Protocol *de facto* a no-first use agreement among states parties during times of peace.

Not surprisingly then, the Geneva Protocol did not act as a deterrent for one state party – Italy – and one signatory state – Japan – to use CW as a means of warfare in the mid- to late-1930s. Italy invaded Abyssinia (now Ethiopia) in October 1935 and before the end of the year tear gas grenades were used by Italian troops. These were replaced first by mustard-filled grenades and later on in the war spray-tanks mounted to planes were used to disseminate the chemical warfare agents.[15] Likewise, the Imperial Japanese Army used CW in occupied Manchuria. As one account sums up:

> The Japanese Army used CW after invading China in 1937, conducting an estimated 1,000 to 3,000 attacks. Japan reportedly produced five to seven million munitions containing agents such as phosgene, mustard, lewisite, hydrogen cyanide, and diphenyl cyanarsine. Although Japanese forces used many of these munitions between 1937 and 1945, a considerable amount was abandoned when Japanese forces retreated.[16]

Short of actual CW use, many states during the interwar period invested in the development and production of chemical warfare agents. In the case of Germany this chemical rearmament stood in stark contrast to the obligations undertaken in the Versailles peace treaty signed after World War I. Yet, it was here, where civilian research into a new group of organophosphorous compounds in the context of research on plastic additives and fertilizers first led to the development and production of the first nerve agent, Tabun, in December 1936. This discovery was followed by the synthesis of Sarin in 1939 and Soman in 1944.[17] Although the UK and US military were also working on this new group of chemical warfare agents, by far the largest portion of the chemical armament efforts went into the classical agents. In spite of the considerable build-up of CW stockpiles before World War II, however, the use in battle of chemical warfare agents did not happen. According to Crone, '[g]as was ... a nonevent in World War II'.[18] The absence of chemical warfare is widely attributed to the deterrent effect, which the chemical stockpiles had on all the belligerents. On the part of the Nazi leadership it is reported that Hitler probably objected to CW use, because he himself was a victim of a CW attack during World War I.[19]

2.2.3. Post-World War II developments

After World War II most of the CW arsenals that had been built up before and during the war were destroyed. However, this does not mean that all CW development and production ceased. Quite to the contrary, the victorious powers were keen on exploiting advances in CW development in Germany in the area of the organophosphorous compounds and to integrate the nerve agents Sarin, Soman and Tabun into their arsenals. Civilian work to exploit the new group of toxic organophosphates continued, leading to the development of even more toxic compounds, some of which were introduced as pesticides but then had to be withdrawn again due to their toxicity to man. One of these super-toxic compounds was adopted by the US military and became known as VX chemical warfare agent in the first half of the 1950s.[20]

With respect to the use of chemical warfare agents, the 1960s saw the widespread use of tear gases and defoliants by US troops during the war in Vietnam. Likewise, allegations of Egyptian CW use in Yemen between 1963 and 1976 were at least partially confirmed.[21] As a reaction to these developments international pressure was mounting to control chemical and biological arms. Negotiations, however, were separated and only a ban on biological weapons was agreed in 1972.[22]

Although negotiations for a CW ban were nominally continuing, a number of states were expanding and diversifying their CW arsenals. One of the most important trends in this area concerned the development of binary munitions. The US government under President Ronald Reagan considerably intensified related efforts in the early 1980s. The basic idea behind the development of binary munitions is that two less- or non-toxic precursor chemicals are mixed in the weapon system – be it a grenade, bomb or missile – to form the actual chemical warfare agent only after employment of the weapon, when it is on the way to its target. This reduces the dangers during the production, handling and storage of these weapons considerably. However, from an economic point of view the utilization of the binary concept has additional advantages: to the extent that binary components have a civilian application (in another end-product) it allows the production of much larger quantities, thereby bringing production cost down and increasing profit margins. Yet, these economic gains come at a non-proliferation cost, as production of such precursors for purportedly civilian applications can disguise illicit production of chemical warfare agents.[23]

In terms of actual usage of chemical warfare agents against both enemy combatants and civilians alike the Iraqi regime of Saddam Hussein stands out in the period since World War II. The Iraqi regime during the 1980s used CW against Iranian troops and against the Kurdish minority in northern Iraq. Investigations carried out on these attacks showed that both mustard gas and the nerve agent tabun were used against the Kurdish population within Iraq.

2.3. The science and technology of CW destruction

Chemical warfare agents were produced in large numbers all through the 20th century. In particular after the two World Wars much of these were disposed of by means no longer acceptable by today's environmental standards. Highly volatile agents like phosgene were released into the open air, others were burned in open pits, dumped into the sea or buried in trenches.[24] Over the past decades more environmentally friendly ways to dispose of chemical warfare agents were developed, the '[t]wo main methods of [which] are by high-temperature incineration or by chemical hydrolysis'.[25] The application of these environmentally sound techniques are also required by international treaties like the 1972 *Oslo Convention for the Prevention of Marine Pollution by Dumping of Wastes and Other Matter* and the 1993 Chemical Weapons Convention. The latter one in Part IV of its Verification Annex explicitly prohibits dumping of chemical warfare agents in water, land burial or open pit burning.[26]

As Pearson and Magee have pointed out in their comprehensive review of existing chemical weapon destruction technologies, the technical problems involved are multidimensional. First of all, chemical warfare agents can be found in assembled chemical weapons, they can be kept in bulk storage containers, or they can be found unexpectedly during building or excavation work.[27] Secondly, technical challenges occur already in the safe handling of the different types of munitions and containers before the destruction process of the chemical warfare agent even begins. If agents are filled in munitions, they have to be extracted from the bomb, grenade or other munitions, which will then be left contaminated. This points to the third issue here: not only need the agents themselves be destroyed according to environmental and health and safety standards, but the munitions or containers in which they had been stored, as well as any additional packaging material that might have been in touch with the chemical warfare agent needs to be decontaminated or destroyed safely.

A historical analysis of destruction technologies applied shows that approximately 80 per cent of chemical weapons destruction since 1958 has used high-temperature incineration as the method of choice. The remainder was destroyed either through neutralization or a combination of neutralization and incineration.[28] As Pearson and Magee in the IUPAC study on CW destruction technologies explain:

> Incineration is an inherently attractive approach for the destruction of organic compounds ... [C]hemical warfare agents are combustible and therefore lend themselves to destruction by incineration. The incineration products are far less toxic than the original chemical warfare agents. ... However, organic compounds containing arsenic present additional problems as many inorganic arsenicals are confirmed carcinogens.[29]

This points to the second type of destruction technology. As the authors of the IUPAC study continue:

> Neutralization involves the reaction of the agent, such as nerve agent GB, with sodium hydroxide causing the nerve agent to hydrolize ... The principal disadvantage of the alkaline hydrolysis of nerve agents is the large volume of hydrolysate produced, which can typically result in a five-fold increase in the overall volume.[30]

In addition, some of the compounds resulting from the neutralization process, i.e. thiodiglycol are themselves precursors listed on the

Schedules of the CWC. This raises questions about the irreversibility of the process and is one additional reason why incineration and not neutralization has been and is likely to remain the method of choice for the largest part of CW agent destruction.[31]

2.4. Dual-use of chemical agents and technology and its impact on the control problem

Any effort to control the use of toxic chemicals for offensive military purposes has to take into account the dual-use nature of many toxic chemicals and related equipment and processes. As the above overview of the developments in chemistry and chemical warfare has already indicated, many toxic chemicals, their precursors and equipment used for their production have perfectly legitimate civilian applications.

To give but a few examples, up until today, chlorine and phosgene are used on a large scale as industrial chemicals in a variety of applications. Phosgene for example is used in 'the manufacture of aniline dyes, polycarbonate resins, coal tar, pesticides, isocyanates, polyurethane, and pharmaceuticals'.[32] Current industrial operations utilizing cyanide-based compounds include:

> fumigation of ships, structures, and agricultural crops ... metal treatment operations, blast furnace and coke oven operations, metal ore processing, and photoengraving operations ... production of intermediates in synthesis of ... dyes, pharmaceuticals, and specialty chemicals ... the manufacture of silver and metal polishes, and electroplating solutions.[33]

In addition, hydrogen cyanide is used in some US states as 'the instrument of execution for convicted criminals in prison gas chambers'.[34]

This widespread use of toxic chemicals for legitimate purposes has implications for both the verification of the peaceful applications of toxic chemicals in the context of implementing the provisions of the CW prohibition regime, i.e. in the states participating in the regime, as well as for the transboundary transfer of toxic chemicals to non-regime members, which will be discussed below.

2.5. Trends in the civilian and military application of chemistry in the late 20th and early 21st centuries

As the discussion so far has shown, chemical warfare agents and means for their production are based on long-established, well-known and proven technologies. Thus, a potential proliferator bent on operating a clandestine CW-programme does not necessarily have to look for the latest developments in chemistry to obtain a militarily significant CW-capability. Nevertheless, several developments are taking place in both the civilian and military applications of chemistry which might well change the way we (have to) think about chemical warfare agents and the ways and means to prevent the misuse of toxic chemicals for offensive military purposes.

2.5.1. New chemical warfare agents through combinatorial chemistry?

The first of these developments concerns the advent of combinatorial chemistry, which:

> comprises a set of techniques for creating a multiplicity of compounds and then testing them for activity ... has been adopted by large and small drug discovery companies alike over the past few years.[35]

This leaves traditional sequential drug discovery procedures, which involved multiple time-consuming repetitions of synthesis and testing far behind. Initial research that forms the foundation of this subfield of chemistry was carried out only in the mid-1980s. Since then combinatorial chemistry has not only revolutionized the way drug discovery works, it is also increasingly being used in other areas like for example pesticide development. In addition to the commercial interest in combinatorial chemistry, '[i]t has been the focus of much academic research as well'.[36] New university courses have been set up, and specialized academic journals to provide an outlet for the research on combinatorial chemistry have mushroomed since the mid-1990s.[37]

According to a conservative estimate of the research activities of the US pharmaceutical industry alone, several millions of new compounds are synthesized every year, approximately 50,000 of which are highly toxic. In the search for medical cures these toxic compounds are of little utility. However, they have one of the key requirements, toxicity, of any potential chemical warfare agent. This does not automatically make them a suitable candidate for this purpose, but since the information on these compounds remains with the companies that first synthesized them, this might provide a very useful resource to tap into in any search for new chemical warfare agents.[38] If the utilization of chlorine and phosgene during World War I and the exploitation of research into organophosphorous compounds for the development of nerve agents is any guide in this matter, developments in the area of combinatorial chemistry need to be monitored very closely indeed. This is all the more important as developments in this area are progressing fast: in order to reduce drug development times a 'novel chemogenomics information system called DrugMatrix' was developed by three US companies.[39] This system contains a 2,000 drug reference set and 'models the new entity's probable effects (biological, toxicological, and clinical) ... a process called predictive profiling.' The misuse potential of a system that allows for the identification of new chemical compounds according to their toxicity is obvious. As data mining algorithms become more elaborated,[40] the potential to identify specific toxic effects of chemical compounds and exploit them for malign purposes will increase.

This is all the more the case as combinatorial chemistry interacts with other enabling technologies in the area of drug development and delivery. In particular, scientific and technological advances in functional genomics,[41] robotics,[42] IT,[43] and nanotechnology[44] act as enablers of combinatorial chemistry and high throughput screening, which in turn have become the driving forces in pharmaceutical research and development.[45]

The genomics revolution, in particular progress in functional genomics, i.e. the ability to attribute specific functions to a particular gene, furthers our understanding of fundamental life processes at a molecular level. To mention but a few examples, such research is concerned with allergies and immunology,[46] breathing,[47] sleep,[48] and depression.[49] Clearly, all this work is geared towards a better understanding of disease origins at the genetic level in order to remedy these diseases. However, the use of a knock-out gas in the Moscow theatre hostage situation serves as a powerful reminder that drugs with perfectly legitimate medical applications might be modified and turned to a different use. Although in the Russian case this use was done by state authorities, the spread of technologies and knowledge brings such misuse potential well within the reach of sub-state groups like terrorist organisations.

2.5.2. The changing face of the chemical industry

Additional challenges to the verification of the peaceful applications of toxic chemicals in industry will be posed by two developments in the chemical industry itself. First, there is a clear trend away from the continuous production of large quantities of a chemical in a facility specifically designed for the purpose. Rather, many companies rely more and more on the use of smaller, more versatile production facilities, which can be adapted from the production of a batch of one chemical to another one in a short period of time.[50] Such facilities could easily fall through the cracks of the declaration and inspection system of the Chemical Weapons Convention. Utilization of such batch-production facilities would theoretically enable a potential proliferator to distribute the production of CW precursor chemicals or chemical warfare agents themselves among a number of such facilities to avoid detection.

Secondly, over the last decade a considerable number of traditional chemical firms were broken up and replaced by so-called 'industrial parks'. This poses a potential problem for verification under the CWC as the Convention's definitions that form the basis for the verification measures assume the existence of plant sites – which were prevalent in the late 1980s, when the CWC was negotiated. A good example of this trend is the evolution of the former Hoechst AG near Frankfurt, Germany, into an industrial park with more than 75 international life science and chemical companies, employing more than 22,000 people.[51] In order to maintain an effective and efficient industry verification system under the CWC, developments like these have to be monitored closely, so as to be able to adapt the verification procedures to the changed environment.

2.5.3. Renewed interest in so-called non-lethal CW

Last, but certainly not least, the recently renewed interest in so-called non-lethal CW threatens to undermine the current prohibition regime and calls into question the viability of any future CW control efforts. If there was the need for a wake-up call to raise awareness of this problem, this was most certainly provided by the use of a 'fentanyl-derivative' – as it was called by the Russian authorities – that was used to end the Moscow theatre hostage-taking in fall of 2002.[52] However, this incident represents just the tip of the iceberg, as more states than just Russia are interested in utilizing so-called non-lethal chemical weapons in a number of police and military scenarios other than war. Certainly the US military shows a strong interest in developing this kind of capability.[53]

From a scientific and technical point of view the major problem with so-called NLW lies in the fact that they are not non-lethal, as the Moscow theatre situation has clearly demonstrated: here about 130 of the 830 hostages present in the theatre died of the effects of the gas used. This represents a percentage of approximately 16 per cent. Compared with the lethality of conventional, commonly assumed to be 'lethal' weapons there is no significant difference: the use of firearms in combat have resulted in 35 per cent fatalities, mines 20 per cent and grenades around 10 per cent.[54] World War I chemical warfare agents like chlorine, phosgene and mustard gas, which are prohibited under the Chemical Weapons Convention, have an even lower lethality of around 7 per cent.[55]

Even if truly non-lethal chemical weapons were technically feasible, is it questionable whether their use would have the effect to merely incapacitate temporarily and not lead to the death of those exposed to the agents. Again, the Moscow theatre scenario offers some insights: Russian security forces obviously had orders to shoot the hostage-takers, which were incapacitated by the gas used in the theatre. Although this might have been the best way to ensure that none of the hostage-takers would be able to detonate any of the bombs which some of them had strapped around their bodies, it reveals a central weakness of the argument of proponents of non-lethal CW. These incapacitants are often used in conjunction with lethal military force and in this context act mainly as a force multiplier, and not as a life-saving tool. Exactly the same pattern of 'non-lethal CW' usage occurred during the Vietnam War, in which the US military employed 10 million pounds of the riot control agent CS:[56]

> A post-war analysis of the operational use of CS declassified in 1979 could find no report of its use against non-combatants or to save civilians and concluded that 'the reduction in casualties has not been in enemy or non-combatant personnel but, rather, friendly troops, as a result of using CS to make other fires more effective'.[57]

3. Scientific and technical issues in relation to the CWC and its implementation

Multilateral negotiations on a ban on chemical weapons were first attempted in the 1960s, when chemical and biological weapons were treated together as the object of an international treaty. After a British proposal to separate negotiations on BW and CW was accepted by the then Soviet Union, negotiations on the Biological Weapons Convention progressed speedily and were concluded in 1972. Although Article IX of the BWC contains a normative guidepost to continue negotiations on a CWC in good faith, it took another two decades before the text of the CWC could be finalized by the Geneva-based Conference on Disarmament (CD) in the fall of 1992.

The CWC was opened for signature in January 1993 and entered into force on 29 April 1997. During the intervening four years a Preparatory Committee (PrepCom) and the Provisional Technical Secretariat (PTS) were preparing the ground for entry into force of the Convention. This involved addressing a number of technical issues related to implementing the CWC that had not been solved during negotiations.[58]

Implementation of the CWC on the international level started in May 1997 with the first session of the Conference of the States Parties (CSP) of the newly set up Organization for the Prohibition of Chemical Weapons (OPCW). Over the course of the first six years of CWC implementation a number of scientific and technological issues related to implementation had to be addressed by both OPCW and states parties to the Convention. When the First CWC Review Conference convened in April 2003, its mandate as contained in CWC Article VIII, paragraph 22 tasked it to 'take into account any relevant scientific and technological developments'.

3.1. Scope of the Convention and its schedules on chemicals

In order to define which acts are permitted and which are prohibited under the Chemical Weapons Convention, a definition of what constitutes a chemical weapon is required. Besides, the drafters of the CWC had to address the question of which toxic chemicals to cover in the new international treaty.

The definition eventually agreed upon in Article II of the CWC distinguishes between 'toxic chemicals and their precursors, except where intended for purposes not prohibited under this Convention, as long as the types and quantities are consistent with such purposes', and 'munitions and devices, specifically designed to cause death or other harm through the toxic properties of those toxic chemicals, which would be released as a result of the employment of such munitions and devices'. As mentioned above, the CWC definition of a chemical weapon does not cover toxic chemicals used as chemical warfare agents against plants. A corresponding clause would not have found the consent of the United States during negotiations of the CWC, as it would have put US use of defoliants like Agent Orange – which in essence is a synthetic analogue of a plant bioregulator – on the spot.

The Annex on Chemicals to the CWC lists the toxic chemicals and their precursors that are considered a risk to the Convention. This Annex divides the substances into three Schedules or lists. Schedule 1 chemicals pose the highest risk to the Convention; many have been developed, produced, stockpiled or used as chemical weapons in the past and they have few if any peaceful uses. Schedule 2 chemicals pose a significant risk to the Convention either because they can be used themselves as chemical weapons or as a consequence of their role as precursors to Schedule 1 or 2 chemicals. Schedule 2 chemicals are also not produced commercially on a large scale. Schedule 3 chemicals are produced in large quantities commercially but pose a risk to the Convention because of their role as precursors to either Schedule 1 or Schedule 2 chemicals.

It has to be emphasized, however, that these three Schedules are neither intended to serve as an alternative definition of a chemical weapon under the CWC, nor are the lists set in concrete. Rather, these lists are used for declarations by states parties and verification activities of the OPCW with respect to routine inspections and as such do not circumscribe the scope of the Convention. It follows from this 'that any chemical, whether listed under a Schedule or not, has to be considered a chemical weapon if it has been produced, stored or used for that purpose'.[59] If required, Schedules can be updated more easily than the text of the Convention itself, in order to reflect changes in the S&T environment.[60]

Riot control agents (RCA) are accorded a special status in the CWC. On one hand, it is acknowledged that they are toxic chemicals, which in principle fall under the prohibitions of the Convention. On the other hand, RCA have legitimate uses in law enforcement and riot control operations. Thus Article I, para 5 prohibits the use of RCA as a method of warfare, while Article II, para 9(d) explicitly permits their use for 'law enforcement including domestic riot control'. RCA are defined as 'any chemical not listed in a Schedule, which can produce rapidly in humans sensory irritation or disabling physical effects which disappear within a short time following termination of exposure' (Art. II, para 7).

Clearly, the recent interest in so-called non-lethal CW or chemical incapacitants is not only confined to the US or Russia and aims at developing more effective and potent RCAs:

> Several countries are currently developing and implementing new non-lethal capabilities. They do so in the belief that this will enable many emerging and non-traditional threats (which may appear in low intensity, asymmetric conflicts and non-combatant operations) to be countered with a progressive response. Non-lethal weapons are of interest to the military and to law enforcement agencies as, in many cases, the character of the scenarios is similar.[61]

Yet, both the organization of the US research and development efforts which have become public and the munitions which are being developed – and in some cases are already on offer – for the dissemination of RCA, raise serious questions about their compatibility with the stipulations of the CWC. Most of the US non-lethal weapons programme is conducted by the US military[62] and the ammunitions envisaged by both US authorities and a Russian weapons producer 'include hand grenades, projectiles for portable grenade launchers, mortar shells and cluster bomb units'.[63]

3.2. CW destruction and its verification

Article I of the CWC obliges states parties to destroy any chemical weapon stockpiles in its possession or which it has abandoned on the territory of another state party. As well, a state party must destroy its chemical weapons production facilities (CWPFs) or convert them for peaceful purposes not prohibited under the Convention. A state party is required within 30 days of ratifying or acceding to the Convention to declare to the Technical Secretariat whether or not it possesses any chemical weapons or has possessed or produced them in the past. Likewise, old chemical weapons (OCW), both those produced before 1925, and those produced between 1925 and 1946, have to be declared. A state party must also notify the OPCW if it has abandoned CW on the territory of another state or if another state has abandoned CW on its territory. These declarations have to include the location and status of any CWPFs and the state party's plan to destroy or convert the facilities. Furthermore, Articles IV and V provide for on-site inspection and monitoring of all locations at which chemical weapons are stored or destroyed.

Chemical weapons must be destroyed within 10 years of the EIF of the Convention, i.e. by 29 April 2007, and the destruction process is to begin within two years of the Convention entering into force. Destruction or conversion activities at CWPFs must begin within one year of the Convention's entry into force, and equally be completed within 10 years. However, a state party may request an extension of up to five years, i.e. until 2012, of the deadline for the destruction of its CW stockpiles.

Six states parties – the United States, Russia, India and South Korea immediately after the CWC's entry into force[64] plus Albania after it had discovered CW on its territory and Libya after its accession to the CWC in January 2004 – have declared the possession of chemical weapons stockpiles. These countries have declared a total of around 70,000 metric tons of chemical agents and about 8.6 Mill. munitions and containers.[65] Eleven states parties have declared a total of 61 current or past CWPFs,[66] nine states parties have declared possessing old CW,[67] and three have declared the existence of abandoned chemical weapons on their territory.[68] Japan has declared that it had abandoned CW on Chinese territory.[69]

In addition, the CWC stipulates intermediate destruction deadlines at certain intervals after entry into force. It soon became evident that Russia would not achieve the first of these intermediate deadlines in April 2000. When applying for an extension, Russia expressed its intention to catch up in its destruction schedule by the next deadline in April 2002.[70] However, this goal also was not realized and Russia had again to apply for extending not only the intermediate, but also the final destruction deadline. In essence, the Russian destruction programme achieved the first destruction goal of 1 per cent of its CW stockpile five years after entry into force, when already 20 per cent of CW stocks should have been destroyed.[71] What is more, the deadline for the complete destruction of the Russian CW arsenal in this second request for extending the deadlines slipped from 2007 to 2012. Since then more requests for extending intermediate destruction deadlines have been submitted to the OPCW by the two late-comer CW possessors Albania and Libya. In both cases, however, the size of the CW stockpiles to be destroyed is small and thus the 2007 deadline for complete destruction of all stocks could be retained.

3.3. Verifying the permitted uses of toxic chemicals and related facilities

Given the above mentioned impact of past developments in chemical technology and industry on military CW production programmes, verifying the permitted uses of toxic chemicals and related facilities had to assume an important role in the overall verification system of the CWC. Activities not prohibited under the CWC are dealt with in Article VI of the Convention and in Parts VI to IX of the CWC's Verification Annex. While the first three of these parts are informed by the subdivision of toxic chemicals into Schedules 1 to 3, Part IX of the Verification Annex deals with other, unlisted, chemicals – so-called discrete organic chemicals or DOCs – and other chemical production facilities (OCPF), which might be easily adaptable to CW production.

As a general rule, and in line with the above mentioned risk assessment for toxic chemicals, Schedule 1 inspections are the most intrusive, most frequent and longest lasting, while intrusiveness, frequency and duration decrease for Schedule 2 and even more so for Schedule 3 inspections. Data monitoring and subsequent on-site verification of a selection of all DOC-facilities is still weaker and falls even more into the category of transparency and confidence-building exercise, 'which – over a number of years – would provide an increasing level of assurance that "CW-capable" facilities were not being used for CW purposes'.[72] DOC inspections are focused in the sense that they first and foremost are applied to OCPFs, which produce chemical compounds containing phosphorous, sulphur or fluorine. 'The underlying assumption was that, with a certain probability, those plants are technologically and chemically closer to Schedule 1 production than others.'[73] Yet, due to the lower overall danger flowing from these types of plants, their inspections were set to begin only three years after entry into force of the Convention, that is in spring 2000.

Between the start of DOC-inspections and the First CWC Review Conference some 100 on-site inspections of OCPFs took place.

Although this represents only a small fraction of the approximately 4,000 inspectable DOC-producing OCPFs, which had been declared by 58 states parties since entry into force, from the point of view of the Technical Secretariat these inspections have:

> shown that there are ... some [OCPFs] that are highly relevant to the object and purpose of the Convention. These facilities produce chemicals that are structurally related to Schedule 1 chemicals. Of particular relevance to the Convention are facilities that combine this kind of chemistry with production equipment and other hardware designed to provide flexibility and containment.[74]

The recognition of these new developments in the chemical industry lies at the heart of calls for an adaptation of the industry verification regime. However, it became clear during the 2003 CWC Review Conference that this assessment is not universally shared by states parties when Pakistan demanded that an '[i]ncrease in emphasis on verification ... of facilities producing relatively harmless discrete organic chemicals (DOCs) should not be at the expense of higher risk Schedule 1, 2 and 3 chemicals listed in the Annex to the CWC'.[75] In the Review Document, the Conference – following an explicit reference to the work undertaken by the OPCW's Scientific Advisory Board – confirmed the 'need to ensure an adequate inspection frequency and intensity' for each category of Article VI facilities. This – according to proponents of the redirection of industry inspection towards the group of OCPF that pose a significant risk to the objects and purposes of the Convention – should allow the necessary measures to be taken by the Executive Council and Technical Secretariat.

3.4. Controlling the transboundary transfer of listed chemicals

Within the confines of the CW prohibition regime, that is among states parties to the CWC, transfers of toxic chemicals have to be declared in relation to Schedule 1 chemicals only. According to Part VI such transfers are permissible 'only for research, medical, pharmaceutical or protective purposes'. Individual transfers have to be notified to the OPCW Technical Secretariat 30 days before they are scheduled to take place by both sender and recipient. In addition, '[e]ach State Party shall make a detailed annual declaration regarding transfers during the previous year'.

During the early phase of CWC implementation it became evident that most transfers of Schedule 1 chemicals among states parties concerned only minute quantities. This led to vast discrepancies in the amounts declared by supplying and receiving states, which in turn resulted in irreconcilable material balances with which the OPCW found itself confronted. Many of the transfers concerned saxitoxin, which is contained in milligram-amounts in ready-packaged kits for the detection of shellfish-poisoning. This technical non-compliance was addressed by raising the declaration threshold for such transfers, which in fact amounted to the first amendment of the CWC.

Trade with non-states parties is regulated according to the chemicals' placement on one of the remaining two Schedules as Schedule 1 chemicals must not be traded with non-states parties. For Schedule 2 chemicals Part VII of the CWC Verification Annex details that three years after the Convention's entry into force all such transfers have to cease. During the interim period the transferring state party has to 'ensure that the transferred chemicals shall only be used for purposes not prohibited under this Convention'. In order to do so, each state party is requested to demand an end-use certificate from the recipient state. Similarly, for Schedule 3 chemicals the CWC contains a grace period for transfers to non-states parties of five years, after which CWC states parties had to decide on how to progress in this regard.

In terms of practical implementation two issues are noteworthy, as they illustrate that industrial concerns are at least as high on states parties' political agendas as the non-proliferation goals of the CWC. First, in relation to Schedule 2 chemicals some states parties – reportedly the US, but also the UK – did not take the requirement for end-user certificates by the competent government authorities in the recipient states very seriously, which led to complaints by those states who did and whose industries felt they were being put at a disadvantage by this uneven implementation.[76]

Second, in relation to Schedule 3 chemicals, states parties had to decide five years after the CWC's entry into force on how to treat trade in these chemicals with non-states parties. In the end, they did not agree on any measures that would have curtailed their industry's interests or hindered S&T advances. The 6th Session of the CSP merely decided to impose a requirement for end-use declarations by states not party to the CWC if the transferred chemical mixture contains 30 per cent or more of a Schedule 3 chemical.[77]

3.5. Review of relevant scientific and technological developments by the First Review Conference

As mentioned above, one of the specific requirements of the First CWC Review Conference was to 'take into account any relevant scientific and technological developments'. Already in the run-up to the Conference a number of contributions in this regard were made by NGOs, including the International Union of Pure and Applied Chemistry (IUPAC), which were then taken up by organs of the OPCW and states parties individually.

The IUPAC report is based on an expert meeting from 30 June to 3 July 2002. Participants in the workshop discussed 'recent technical developments and ... the state of the art in several areas of organic synthesis, industrial chemical processing, and analytical chemistry methodologies', and identified five 'key scientific and technological areas that should be taken into account at the First Review Conference'.[78] These are technical challenges to the Convention, analytical techniques for routine and challenge inspections as well as investigations of alleged use of CW, technical capability of the TS, education and outreach, and issues related to CW-destruction.

The Scientific Advisory Board took up many of these themes in its report that was submitted to the Review Conference. The SAB *inter alia* noted that it:

> was aware of concerns about the development of new riot control agents (RCAs), and other so-called 'non-lethal' weapons utilising certain toxic chemicals (such as incapacitants, calmatives, vomiting agents, and the like). ... The SAB noted that the science related to such agents is rapidly evolving, and that results of current programmes to develop such 'non-lethal' agents should be monitored and assessed in terms of their relevance to the Convention. However, based on past experience and the fact that many of these compounds act on the central nervous system, it appears unlikely from a scientific point of view that compounds with a sufficient safety ratio would be found. ...
>
> The SAB stressed the importance that all new toxic chemicals, no matter what their origin or method of synthesis, are covered by the Convention's definition of CW, unless they were intended for purposes not prohibited by it, and only as long as their types and quantities would be consistent with these purposes. The SAB

underlined the importance of this aspect of the definition of CW as a safeguard for the validity of the Convention.[79]

With the opening of the Conference, however, S&T issues in general disappeared almost completely from sight, as they were fed into a number of components of the review of the CWC's operation. However, S&T issues – more specifically the Report of the SAB as submitted to the Conference by the Director General – resurfaced in the Review Document both in the sections on general verification provisions and on activities not prohibited under the CWC.

One S&T issue which received considerable attention in the run-up to the meeting and to which the above quote from the SAB report explicitly refers, was almost completely suppressed during the Conference: the question of chemical incapacitants and so-called 'non-lethal chemical weapons', in which some CWC states parties recently have shown a renewed interest. Although two states parties – New Zealand and Switzerland – made explicit reference to non-lethal weapons during the General Debate, the International Committee of the Red Cross (ICRC), whose statement was focusing on chemical incapacitants, was not allowed to address the plenary. As a result, the only opportunity to discuss these matters publicly arose with the *Open Forum on the Chemical Weapons Convention*, hosted by the TS and supported by a number of NGOs. The *Open Forum* included a panel discussion on 'The Chemical Weapons Ban and the Use of Incapacitants in Warfare and Law Enforcement'.[80]

Not surprisingly, then, informal discussions among delegations showed that the time was not ripe for the inclusion of any language explicitly referring to incapacitants or non-lethal weapons in the text of the Review Document. However, the document did contain language in relation to the definitions in Article II of the Convention, pointing out that these were found by the Conference to adequately cover developments in science and technology.[81] In addition, the Conference tasked the Council to consider developments in relation to new chemicals that may be relevant to the CWC.

4. Summary and conclusions

As the overview in the first substantive part of this chapter has shown, the way in which chemistry and the industrial applications of chemical technology developed since the late 19th/early 20th century had a profound impact on the possibilities for waging chemical warfare. Only when the liquefaction of chlorine and its storage in pressurized cylinders became possible on an industrial scale, did this new technology provide enough chemical warfare agents to the military planners of the day to warrant their employment on the battlefields of World War I. Similarly, civilian work – first in the 1930s and then in the 1950s – on highly toxic organophosphorous compounds paved the way for the development of a new class of chemical warfare agents – the nerve gases. The lesson to be learned from this is that when opportunity arose to exploit toxic chemicals in the past, the military planners in a large number of countries were quick to utilize these opportunities.

In parallel to these advances in chemical science and technology and their industrial applications, the chemical weapons prohibition regime evolved. The major part of regime development took place in the second half of the 1980s when the negotiations for the CWC were gathering steam. It is almost inevitable that the drafters of the CWC, when devising the mechanisms to verify compliance with the CWC's obligations, did so on the basis of the chemical industry of that time.

This leads to a situation today in which the regime structure is addressing problems of the past – or the present, at best – and needs to be reoriented and supplemented by additional measures to meet future threats, which again can be assumed to mirror scientific and technological developments in chemistry and the life sciences more broadly. Clearly, the CW control regime contains procedures to address S&T change and an organizational structure in the form of the Scientific Advisory Board to utilize scientific expertise to address such change. Yet, the manner in which these mechanisms have been used by states parties and the way in which existing obligations have been implemented does not bode well for states parties' future willingness and capacities to adapt to the technological challenges ahead. As evidenced by the debate on DOC-inspections and the suppression of the question of chemical incapacitants, most states parties place a higher priority on safeguarding what they obviously perceive as national interests than they do on making the regime 'future-proof'.

3

Science, Technology and the BW Prohibition Regime

1. Introduction

The biological weapons (BW) prohibition regime is built around the 1972 Biological and Toxin Weapons Convention (BWC), the 1925 Geneva Protocol, and the Australia Group, which expanded its activities from CW-related dual-use goods and technologies into the BW realm in 1990.

As in the case of the CW prohibition regime, scientific and technological developments in the life sciences determine the options available to wage biological warfare and also have a profound impact on the effectiveness of efforts to prohibit BW and the robustness of the regime created for this purpose. Therefore the first section of this chapter addresses the technical characteristics of biological warfare agents and related technologies, and outlines the dual-use character of some of these. The following section introduces the BW prohibition regime and its development through the activities of subsequent BWC Review Conferences, and the Australia Group. The attempts to strengthen the regime through the negotiation of a legally binding protocol to the BWC are then dealt with. The final section analyses the development of the BW prohibition regime since the 5th Review Conference and asks to what extent the regime in its current form is able to contain the dangers emanating from the new scientific and technological developments in the life sciences.

2. Technical characteristics and the dual-use problem of BW agents and related technology

2.1. Developments in biology and biological warfare during the 20th century

The use of biological agents in warfare goes back at least several hundred years.[1] However, only with the advances in the scientific understanding of life and its underlying processes towards the end of the 19th century has a systematic utilization of pathogens and naturally produced toxic substances for warfare purposes become possible.

Biological warfare agents are usually grouped into five categories: 1) bacteria, such as *Bacillus anthracis*, the causative agent of anthrax, *Yersinia pestis* which causes bubonic plague, and *Francisella tularensis* which causes tularemia; 2) viruses, such as the ones that cause smallpox, Ebola and Venezuelan equine encephalitis; 3) rickettsiae which can cause Q-fever and typhus; 4) fungi, such as the *Aspergillus* fungi; and 5) toxins, which are non-living products from micro-organisms but also plants or animals, like botulinum toxin, ricin, or saxitoxin, respectively. So, most biological warfare agents are not only different from CW agents,[2] but it is a diverse group in and of itself, in which some agents are mostly incapacitating, while others have a rather high lethality. Also, some biological warfare agents will be localized in their effects while others – due to their contagiousness – may cause widespread epidemics. Following from this diversity, biological warfare agents can be employed in a number of attack scenarios, ranging from assassinations of individuals to large-scale aerosol releases, which could theoretically cover whole population centres and affect many thousand people.[3]

While the often observable focus on pathogens that would be employed against humans is understandable, one must not lose sight of a number of animal and plant diseases, whose causative agents could be utilized for BW attacks against crops or livestock.[4]

To make control efforts even more complicated, the material, technologies, and know-how needed for offensive military BW programmes or the pursuit of terrorist BW attacks are of a so-called dual-use character. Not only can they be used for offensive purposes, but many of the 'ingredients' of a BW programme have perfectly legitimate applications. Thus, it cannot be deduced from the mere presence of a seed culture of a particular pathogen or a specific type of equipment that a state pursues an offensive BW programme. This material and equipment might be employed in a perfectly legitimate civilian use such as the production of dairy products or vaccine production. Alternatively, the state might be pursuing some military biodefence activity – which is permitted under the provisions of the BWC.

Furthermore, the nature and scope of biological warfare has changed dramatically by the revolution in the life sciences. As has been shown for the 'three generations of offensive biological warfare programs' of the 20th century, all the military programmes were 'developing on the back of growth in scientific knowledge'.[5] According to this account, military BW programmes followed scientific discoveries in the areas of 1) bacteriology, laying the ground for the BW-based sabotage activities during World War I, 2) aerobiology, providing for the knowledge to spread biological warfare agents over large geographic areas, and thereby giving also non-contagious agents their potential to be used as mass casualty weapons, and 3) genetic engineering, which played an important role in the offensive BW programme of the former Soviet Union. Following the revelations of one former high-ranking programme insider, one has to assume that the Soviet programme included work on and the weaponization of genetically modified pathogens such as 'antibiotic-resistant strains of plague, anthrax, tularemia, and glanders'.[6]

2.2. Ascent of the biotech-industries in the last quarter of the 20th century

The scientific and technological advances on which the offensive Soviet BW programme could build, date back to the early 1950s when the ladder-like double helix structure of DNA was discovered by James Watson and Francis Crick. This revelation was complemented by the discovery of the first restriction enzyme by Daniel Nathans, which can cut genetic material into pieces for various studies and applications. In 1973 Stanley Cohen and Herbert Boyer 'developed a laboratory process for joining and replicating DNA from different species'[7] called recombinant DNA technology, and widely considered to mark the birth of modern biotechnology. They completed the first successful genetic engineering experiment by inserting a gene from an African clawed toad into bacterial DNA. In 1977 William Rutter and Howard Goodman isolated the gene for rat insulin and transplanted it into bacteria. Five years later, genetically engineered human insulin went onto the market.

While most of the basic research in the beginning of the biotechnology age was done at university laboratories, most notably Stanford University and the University of California at San Francisco, the academic pioneers of biotechnology soon set up their own companies to exploit these scientific advances commercially.[8] In the US, where the biotechnology industry is further developed than the rest of the world, it was dominated in its early years by a handful of US firms including Amgen, Biogen and Genentech. During the late 1980s these were the only companies that appeared to have products on the market that were earning substantial profits. This was largely due to the huge commercial success of two drugs, Epogen and Neupogen produced by one company, Amgen.

Since then, the number of new drugs based on biotechnological production methods has increased dramatically, as has the number of companies in the biotechnology sector. While only a few start-ups populated the commercial biotechnology scene in the late 1970s and early 1980s, by the end of the century most major pharmaceutical companies had moved into the area and the worldwide number of biotech companies had grown to several thousand:

> Less than 30 years ago the industry began with a handful of US start-ups using genetic engineering to manufacture commercial quantities of well-characterized human protein drugs. Today the global biotech industry includes more than 4,000 companies throughout the US, Canada, Europe, Australia/New Zealand and Asia applying revolutionary discovery science to tackle the planet's toughest healthcare, agricultural, industrial and environmental challenges.[9]

With the large number of new processes and applications in the biotech field have also come more opportunities for misuse of these advances for malign purposes.[10] As a panel of life sciences experts concluded in a recent assessment of the threat of advanced BW based on biotechnological methods and processes that was conducted for the CIA:

> ... other classes of unconventional pathogens that may arise in the next decade and beyond include binary BW agents that only

> become effective when two components are combined [...]; 'designer' BW agents created to be antibiotic resistant or to evade an immune response; weaponized gene therapy vectors that effect permanent change in the victim's genetic make up; or a 'stealth' virus, which could lie dormant inside the victim for an extended period before being triggered.[11]

Awareness of the biotechnology sector of these potential misuses of their work has increased since the anthrax attacks in the US in the fall of 2001. As the 2002 Ernst & Young report notes:

> The biotechnology industry's roots lie in the peaceful applications of this revolutionary technology. But in the post-11 September world, it finds itself having to consider mixing its normal business with national defence efforts. Whether it is in assisting in the development of vaccines for infectious diseases that may never be unleashed, reducing the cost of medicines to battle public health emergencies, or complying with new restrictions on government inspections of laboratories and manufacturing plants, the life sciences sector is now on the front lines of the defence against terrorism.[12]

Although the realization seems to have sunk in with parts of the life sciences sector that their services might be required in the fight against BW, setting up an incentive structure for the biotechnology industry to actually get involved in the development of vaccines or other medical countermeasures is still in its early stages, but gaining momentum in the states which are most advanced in this regard. In the United States for example 'Project Bioshield' aims to encourage private industry to develop countermeasures sought by the government in its fight against terrorism, but for which there is no commercial market, by providing assurances to life sciences and other companies that the products they develop will actually be purchased by the US government. 'Project Bioshield' was initiated in President Bush's 2003 State of the Union speech and in its current form (S.15), as approved by the US Senate the bill authorizes $5.6 billion over 10 years for the government to procure drugs, biological products and devices as countermeasures against biological, chemical, radiological and nuclear agents that could cause a public health threat.[13]

In addition to creating such an incentive structure for the biotech industry, the Project Bioshield Act of 2004 also allows for expedited peer review procedures for research and development of bioterrorism countermeasures and allows the Secretary of Health and Human Services together with the Food and Drug Administration (FDA) to permit the distribution of pharmaceuticals and other products for emergency use prior to FDA approval for general use.

3. Setting-up and developing the BW prohibition regime

The origins of the BW prohibition regime date back to the 1925 *Protocol for the Prohibition of the Use in War of Asphyxiating, Poisonous, or Other Gases, and of Bacteriological Methods of Warfare.*[14] The Protocol was originally conceived as a response to the widespread use of chemical weapons (CW) during World War I, and only upon a Polish initiative were 'bacteriological methods of warfare' included into the Protocol text.[15] It entered into force in 1928 and has currently 133 member states.[16] Upon ratification or accession to the Geneva Protocol many states issued unilateral declarations laying down restrictions under which they would themselves consider to be bound by the provisions of the Protocol.[17] As a result, it has been widely regarded as an agreement that provides for no-first use of CBW among states parties to the Protocol. This assessment had to be modified in recent years as a number of states have given up their reservations.[18]

3.1. BWC negotiations and content

During most of the 1960s states which sought to improve controls on chemical and biological weapons were advocating one of two courses of action: either in the context of proposals for 'General and Complete Disarmament', where at some point on the way to achieving this goal, CBW would have to be addressed,[19] or through strengthening the 1925 Geneva Protocol. In 1968 the UK submitted a working paper to the ENDC in which it deviated from this pattern and proposed treating BW separately from CW.[20] Although some initial resistance had to be overcome, the proposal, after the US unilateral renunciation of BW and toxin weapons in 1969 and 1970,[21] and Soviet acceptance of the idea of separating negotiations on BW and CW, led in a short period of time to the conclusion of the 1972 Biological Weapons Convention.

The BWC is based on the recognition that the use of BW agents constitutes an abhorrent act of warfare and is therefore prohibited. This principle, which is also referred to as the 'BW taboo' is explicitly mentioned in preambular paragraphs 9 and 10, in which states parties to the BWC express their determination:

> for the sake of all mankind, to exclude completely the possibility of bacteriological (biological) agents and toxins being used as weapons, convinced that such use would be repugnant to the conscience of mankind.

At the same time, peaceful uses of the biosciences are regarded as a legitimate undertaking. According to BWC Article I:

> Each State Party to this Convention undertakes never in any circumstances to develop, produce, stockpile or otherwise acquire or retain: (1) Microbial or other biological agents, or toxins whatever their origin or method of production, *of types and in quantities that have no justification for prophylactic, protective or other peaceful purposes.* (emphasis added)

This so-called general-purpose criterion not only makes it clear that peaceful uses of the biosciences are legitimate undertakings for states parties to the BWC, but also allows the use of pathogenic organisms or toxins in quantities and for purposes other than use as weapons.

From this reasoning, it can be inferred that states subscribing to the regime regard defences against the threat or use of BW as permitted. This principle is rooted in the belief that the peaceful uses of biosciences cannot be taken for granted – be it for the lacking universality in membership or for a state party not living up to the obligations it has assumed.

As the above quoted Article I makes clear, five activities related to BW – development, production, acquisition by other means, stockpiling, retention – are explicitly banned. Yet, the scope of the BW prohibition regime is wider than these five activities and the treaty contains several more normative guidelines for state action. Central to the BW prohibition regime is the non-use norm, which is explicitly spelled out in the 1925 Geneva Protocol and implicitly contained in Article I of the BWC. Although this article of the Convention makes explicit reference only to the non-acquisition norm, it can be inferred from this that use is prohibited as well.

The disarmament norm is contained in Article II of the BWC. It requires that all states parties either destroy or divert to peaceful purposes all agents, toxins, equipment and means of delivery related to their BW holdings within nine month after entry into force of the BWC. Unilateral declarations that this obligation had been complied with by the three BWC depositary states (UK, US and Soviet Union) were made in March and June 1975, respectively.[22]

According to the non-transfer norm, which is contained in BWC Article III, states parties forswear to 'transfer to any recipient whatsoever, directly or indirectly, and not in any way to assist, encourage, or induce' any actor to acquire any of the items specified in Article I of the BWC. The non-transfer norm is strengthened through Article IV of the BWC, which calls for national implementation measures to put the basic obligations under the Convention into effect.

In addition, Article X of the BWC contains the cooperation norm, which, from the point of view of some BWC states parties from the developing world represents the flip side of the deal's non-acquisition and non-transfer norms. Article VII contains the assistance norm, according to which states parties will come to each other's assistance in case of the use or threat of BW against one of them. The consultation norm is spelled out in Article V of the BWC, in which states parties agree to 'consult one another and to co-operate in solving any problems which may arise in relation to the objective of, or in the application of the provisions of, the Convention'.

In addition, the continuing link between BW and CW disarmament is acknowledged through the normative requirement for BWC states parties to continue negotiating a CW treaty, as spelled out in Article IX of the BWC.

Furthermore, the harmonization norm, although operationalized in a somewhat different form, guides the behaviour of states participating in the Australia Group, which have agreed to harmonize their export control policies, to share information on 'suspicious' requests for supplying CBW-related dual-use items and technologies, and to consult one another in case of export denials of certain dual-use items and technologies to states of proliferation concern.

3.2. Developing the regime: BWC Review Conferences and the Australia Group

The two central weaknesses of the BW prohibition regime – the absence of a verification principle and precise rules and procedures that would specify how to implement the norms of the regime in everyday state practice – came to the fore soon after entry into force of the BWC in 1975. Already during the First BWC Review Conference in 1980 the US voiced its concern that an anthrax outbreak in the then Soviet city of Sverdlovsk had been caused by a clandestine military facility, which, it was suspected, was part of an offensive Soviet BW programme.[23] Although the Soviet Union denied the allegations, doubts lingered and during the following Second and Third BWC Review Conferences in 1986 and 1991, respectively, a number of confidence-building measures (CBM) were agreed upon.

3.2.1. Scope of the BWC and advances in science and technology

As discussed above, the scope of the BWC is outlined in its Article I, which contains the explicit prohibitions. Implicitly, these five prohibitions also prevent states parties from using BW agents. What is not prohibited, however, is research involving BW agents or defensive measures against BW which might also involve production of live agents and their testing in aerosol chambers, for example in order to ascertain the usability of a respiratory mask. This points to a second dual-purpose dilemma in controlling BW in addition to the one posed by the dual-use dimension involved in the applicability of many BW-related materials and technologies in the civilian realm.

Although the general purpose criterion contained in Article I of the BWC makes clear that in principle all future scientific and technological advances are covered by the Convention, states parties over the years have regarded it as important to have a statement to this effect included in the Final Declaration of successive Review Conferences. At the First Review Conference in 1980 this reaffirmation of the comprehensive scope of Article I merely stated that:

> The Conference believes that Article I has proved sufficiently comprehensive to have covered recent scientific and technological developments relevant to the Convention.[24]

The brevity of this statement is not surprising as the biotechnology revolution was still in its infancy at that point in time. With advances in biotechnology and genetic engineering steadily progressing, the Second Review Conference in 1986 in its final declaration already saw the need to be somewhat more specific in mentioning the fields of S&T progress which states parties were most concerned about being misused. The resulting passage in the 1986 Final Declaration said that:

> The Conference, conscious of apprehensions arising from relevant scientific and technological developments, inter alia, in the fields of microbiology, genetic engineering and biotechnology, and the possibilities of their use for purposes inconsistent with the objectives and the provisions of the Convention, reaffirms that the undertaking given by the States Parties in Article I applies to all such developments.
>
> The Conference reaffirms that the Convention unequivocally applies to all natural or artificially created microbial or other biological agents or toxins whatever their origin or method of production. Consequently, toxins (both proteinaceous and non-proteinaceous) of a microbial, animal or vegetable nature and their synthetically produced analogues are covered.[25]

The final declaration of the Third Review Conference in 1991 basically repeated the 1986 statement concerning scientific and technological advances of relevance to the BWC. States parties at the Fourth Review of the BWC in 1996, however, felt the need to add to the previous statement by pointing out that 'any application from genome studies' was covered by the BWC's prohibitions as well.[26] Thus, the states parties proved to be very perceptive of future applications of scientific breakthroughs and included genome studies applications well before the human genome actually had been decoded.

The continuous and accelerating progress in various areas of the life sciences between the Fourth and the Fifth Review Conferences in 2001 were reflected in a number of submissions by states parties to the Fifth Review Conference. As the US statement noted:

> Since the 4th Review Conference in 1996, there have been significant advances in the field of biotechnology. The major advances have occurred in the fields of genetic modification, genomics, proteomics, bioremediation, biocontrol agents, vaccine development

> and bioinformatics. Of special interest to the BWC are applications in directed molecular evolution (i.e., genetic modification), proteomics, bioinformatics, and vaccinology. The number of countries which are developing and enhancing their biotechnology capabilities continues to grow as the applications continue to expand into commercial sectors and the resulting industry has expanded in both scope and products developed and marketed. All of these trends continue to have practical significance for the BWC.[27]

Whereas the UK and the Swedish background papers were equally broad in scope and tried to cover all aspects of S&T developments of the previous five years that are of relevance to the BWC, South Africa focused 'exclusively on developments in terms of biocontrol agents and plant inoculants',[28] thereby reminding states parties that the prohibitions of the BWC apply to biological warfare against plants – and animals, for that matter – as well. Unfortunately, due to the failure to successfully conclude the Fifth Review Conference and the concomitant absence of a Final Document,[29] the shared interpretations of BWC states parties concerning scientific advances of relevance to the BWC have not been recorded in a consensual document.

3.2.2. From non-verification to confidence-building measures

The seeds for the confidence-building measures (CBMs) that were agreed upon in 1986 and 1991 were sown during the first BWC Review Conference in 1980. The Final Declaration of that Conference noted in relation to Article II that voluntary declarations by state parties on the destruction of their former BW stockpiles 'contribute to increased confidence in the Convention'.[30] A paper delivered by a US diplomat in 1984 contained the first elements of the transparency measures, which would later become the CBMs.[31]

The confidence-building measures (CBMs) agreed upon by the Second and Third Review Conference in 1986 and 1991, respectively, consist of a politically binding commitment of all states parties to participate in annual exchanges of data and information, as well as declarations of past and present activities of relevance to the Convention. More specifically, the CBMs include:

- Measure A, Part 1: Exchange of data on research centres and laboratories that meet very high national or international safety standards (WHO BL4/P4).

- Measure A, Part 2: Exchange of information on national biological defence research and development (R&D) programmes, including declarations of facilities where biological defence R&D programmes are conducted.
- Measure B: Exchange of information on outbreaks of infectious diseases and similar occurrences caused by toxins that seem to deviate from the normal pattern.
- Measure C: Encouragement of publication of results of biological research directly related to the Convention and promotion of use of knowledge.
- Measure D: Active promotion of contacts between scientists, other experts and facilities engaged in biological research directly related to the Convention, including exchanges and visits for joint research on a mutually agreed basis.
- Measure E: Declaration of legislation, regulations and other measures including exports and/or imports of pathogenic micro-organisms in accordance with the Convention.
- Measure F: Declaration of past activities in offensive and/or defensive biological R&D programmes since 1 January 1946.
- Measure G: Declarations on vaccine production facilities, licensed by the State Party for the protection of humans.

The actual submission of these annual declarations by BWC states parties, however, has been unsatisfactory at best. As one of the few publicly available reviews of CBM submission for the 1996 BWC Review Conference shows:

> it has taken nine years of participation to reach the stage at which over half of the States Parties to the BWC [75 out of 138] have made at least one annual declaration. ... Only about one-third of the States Parties to the BWC takes part in the information exchange under the CBMs per year.[32]

That this dismal state of affairs in terms of CBM submission has not changed since the Fourth Review Conference in 1996 is all the more regrettable as some of the measures agreed upon are explicitly related to advances in science and technology and as such could provide at least some reassurance that new scientific developments are not misused in clandestine offensive BW programmes.

3.2.3. Proliferation concerns and export controls on dual-use goods and technologies

Concerns about the proliferation of chemical and biological weapons-related material and technologies were most pronounced during the early to mid-1980s, when evidence was mounting that Iraq had produced and used CW in its war against Iran. As a reaction to these revelations and the involvement of Western dual-use exports, which found their way into the Iraqi CBW programmes, the Australia Group was formed in 1984 in order to harmonize export control policies among supplier states.

Before discussing the Australia Group's activities it might be useful to take a step back and consider the roles and functions of export controls in general terms, for they are not, as often portrayed by their critics, a form of technology denial in disguise. Rather, their primary goal is to ensure the civil application of exported commodities and services and to deter from using them in military programmes. To fulfil this function domestically, that is in the supplying state, export control measures have to be capable of identifying illegal exports and have to threaten to the potential exporter a level of punishment that exceeds any gain from such an illegal export. Although export controls on dual-use goods and equipment pursued in isolation from other non-proliferation measures cannot prevent weapons acquisition by a determined proliferator, they can slow down the procurement process and increase the proliferator's costs. When coordinated among supplier states, export controls provide a level playing field for potential suppliers in different states, thereby increasing the hurdles a proliferator has to take even more. Harmonized export controls make it more difficult to play one supplier against the other, since the individual exporters do not have to face comparative disadvantages because of unequal export control guidelines.[33]

With a view to the BW prohibition regime, their proponents regard export controls as a means to implement the non-transfer commitment of Article III of the BWC and their above mentioned domestic dimension clearly falls into the purview of Article IV of the Convention. Consequently, any argument that export controls – which are devised and implemented with the above functions in mind – contravene the underlying principles of the BW prohibition regime is difficult to sustain.

The Australia Group's original purpose was to constrain the trade in technologies and materials of chemical warfare. An additional, equally important function of the Group is the exchange of intelligence and other information concerning the procurement activities of suspected proliferators. Membership of the Group has more than doubled from its fifteen founding members of two decades ago to now 39 and the group:

> meets annually to discuss ways in which the national level export licensing measures of its ... participants can collectively be made more effective in ensuring would-be proliferators are unable to obtain necessary inputs for CBW programs, which are banned under international law.[34]

In 1990 Australia Group members agreed to expand their controls to cover BW agents and toxins, as well as dual-use equipment necessary for their production. From 1992 onwards control lists were implemented which initially covered eighteen bacteria, four rickettsiae, 25 viruses, and fourteen toxins. In addition, dual-use equipment like fermenters, centrifuges, aerosol chambers, and filter and freeze-drying equipment with certain technical specifications were subjected to controls.[35] The control lists of the Australia Group have been under review and were updated during the 2004 plenary of 7–10 June:

> As part of the Group's ongoing efforts to keep its common control lists up to date and scientifically relevant, participants agreed to add five plant pathogens to the control lists – the first such addition since 1993 – and to expand medical exemptions for one of the controlled toxins. Participants also advanced consideration of further additions to the control lists, including airborne spraying and fogging systems capable of dispersing biological agents in aerosol form.[36]

4. Attempted strengthening of the regime I: VEREX and the Ad-Hoc Group of States Parties

4.1. Determining technical feasibility: the VEREX exercise

In addition to agreeing on an expanded set of CBMs, the 1991 BWC Review Conference established the so-called VEREX group of governmental experts to investigate the technical feasibility of measures to enhance the confidence in compliance with the BWC's prohibitions. This group met four times from 1992 to 1993 and examined 21 potential verification measures from a scientific and technical viewpoint. These measures included off-site activities like information monitoring, data exchange, and remote sensing, as well as on-site measures like exchange visits, inspections and continuous monitoring.

The VEREX group concluded in its report, which it submitted to all BWC states parties:

> that potential verification measures as identified and evaluated could be useful to varying degrees in enhancing confidence, through increased transparency, that States Parties were fulfilling their obligations under the BWC. While it was agreed that reliance could not be placed on any single measure to differentiate conclusively between prohibited and permitted activity and to resolve ambiguities about compliance, it was also agreed that the measures could provide information of varying utility in strengthening the BWC.[37]

Although no single measure was deemed sufficient to provide enough reassurance about treaty compliant behaviour, the approach adopted by VEREX, that is to favour a combination of measures, prompted enough states parties to the BWC to seek a Special Conference (more than half the BWC states parties were required for this). This Special Conference took place in September 1994 and served the critical function of converting the 'VEREX findings, which were exclusively scientific and technical, into the basis for diplomatic efforts. Following the Special Conference diverse political and economic considerations could be introduced in the context of a multilateral negotiating process.'[38]

4.2. Mandate for the negotiations and scope of the Protocol

It was exactly these 'political and economic considerations' which made it difficult for the Special Conference to arrive at a consensual negotiating mandate for the Ad-Hoc Group (AHG) of States Parties to negotiate a legally binding protocol to the BWC. First of all, the US was highly sceptical of the AHG to negotiate 'verification' measures, because it rejected the idea that the BWC is verifiable on principled grounds. That this view did not change over the course of the negotiations was confirmed by one of the main negotiators in the US delegation to the AHG, Ambassador Donald Mahley, in testimony before the US Senate in September 2000. As Mahley explained:

> First of all, this is not an issue of verification. As you know, the United States has substantive requirements for attributing effective verification to a treaty. ... The United States has never, therefore, judged that the Protocol would produce what is to us an effectively verifiable BWC ... [39]

As a result of this US approach to what it considers to be 'effective verification',[40] the AHG was tasked to negotiate 'measures to enhance compliance' with the BWC. Secondly, Russia insisted that the AHG consider 'definitions and objective criteria' in its mandate. Thirdly, a number of states of the Non-Aligned Movement (NAM) did not see the necessity of the whole exercise in the first place. Many of them did not regard BW as a threat to their national security and therefore had to be brought into the process by including into the mandate the negotiation of measures to strengthen international cooperation in the peaceful uses of the biosciences.[41] Lastly, the already existing – but poorly implemented – CBMs should be taken into account by the negotiators of the AHG and new CBMs should be considered.[42]

Negotiations started in January 1995 and progressed until July 2001, when the overall approach taken in the negotiations up to that point was rejected by the US administration and the draft protocol text declared to reduce rather than increase security against BW. The negotiations proceeded in four phases: in the first one, which lasted until mid-1997 potential elements of a compliance protocol were identified. From mid-1997 negotiations were based on a rolling text, which was developed further during the negotiating sessions of the AHG. During the third phase in 1999 the formal structure of the Protocol was negotiated and in the fourth phase compromise language on the less controversial issues under negotiation was integrated in the rolling text.[43] In order to create additional momentum for the AHG to enter into an end-game during which also the more controversial issues could be tackled, the chairman of the AHG developed a compromise text which he presented to delegations in spring of 2001.[44]

4.3. Compliance measures

4.3.1. Declarations

Declarations by states parties on certain facilities and activities of direct relevance to the object and purpose of the BW prohibition regime form the basis of measures to increase the transparency of states parties actual behaviour. Only on the basis of declarations of dual-use industry facilities, which in principle could be easily transformed from permitted to prohibited use, can a coherent system of declaration follow-up procedures including for example visits be established.

In the negotiations of the AHG it became obvious very early on that the sensitivities of some states parties as to a potential loss of national security relevant or commercially sensitive information were much higher than for example in the case of the CW prohibition regime. This higher sensitivity is related to both the character of the biotechnology and pharmaceutical industries – which are not as old as the chemical industry – and with military concerns that too much openness about biodefence activities might reveal one's weaknesses to a potential aggressor contemplating the use of BW.

The approaches favoured by states with biodefence activities clearly sought to minimize the burden that the Protocol would put on their own biodefence programmes and activities through the inclusion of specific declaration triggers. States with big defensive programmes for example favoured that only sites be declared, where more than 15 person years were devoted to R&D activities in work on pathogenicity or virulence, aerobiology or toxicology.[45] In contrast, small biodefence programme operators sought to include language in the Protocol so that biodefence facilities had to be declared where more than five person years or persons are dedicated to biodefence work and list those facilities (without getting into the specifics involved in a declaration) where between two and five persons are involved in biodefence work.[46]

The treatment of declarations in the Compliance Protocol reflects these diverging interests both with respect to initial declarations as well as annual declarations that states parties would have to submit.[47] According to Article 4 of the Protocol, initial declarations of a number of facilities and activities have to be submitted within 180 days after entry into force of the Protocol; annual declarations not later than the end of April for each year.

Initial declarations are required (1) for past offensive BW programmes which BWC states parties acceding to the Protocol have undertaken between 1946 and entry into force of the Convention for the respective state – in case of the USA and the Russian Federation the latter year would be 1975, and (2) defensive BW programmes and/or activities conducted during the ten years preceding the entry into force of the Protocol for the state party.

Article 4 further specifies the requirements for annual declarations of national biological defence programmes and/or activities, maximum biological containment facilities, high biological containment facilities which exceed 100 m^2 and have produced vaccines or other specified production or have carried out genetic modification of any agent or toxin contained in the 'List of Agents and Toxins' as specified in Annex A to the Protocol, plant pathogen containment, specified work with listed agents and toxins and specified production facilities. These provisions detailing the declaration requirements for states parties to the Protocol would have had a positive impact on the effectiveness of the regime, in so far as they would have provided part of the regulatory basis for a new transparency norm. At the same time, the combination of declaration triggers would have led to a situation where only the most relevant facilities – in terms of presenting a danger to the object and purpose of the regime – would fall under the declaration requirement. Although this would not completely eliminate the risks of loss of national security relevant or confidential proprietary information, it clearly would have reduced them.

4.3.2. Visits

Proponents of non-challenge visits see a number of positive functions visits could play. They argue that such on-site measures increase the transparency of BW-relevant activities, enhance confidence in the treaty-abiding behaviour of states parties by increasing the likelihood that violations are detected, contribute to the clarification of ambiguous declarations, and enhance cooperation among states parties. Critics of non-challenge visits counter that the detection of treaty violations through such weak on-site measures is highly unlikely. Rather, there exists – according to this group of states, which includes the US – the danger that visits lead to the loss of confidential business or national security information. Consequently, if they cannot be avoided altogether, the frequency and intrusiveness of such on-site measures has to be minimized.[48]

Over the course of the AHG negotiations, a number of different visit concepts have been advanced and hotly debated. Discussions on these concepts have displayed a second dividing line, which relates to the question of whether visits should only be conducted in declared facilities or whether non-declared ones should be a possible target for visits too. While the Western Group in the AHG hold the latter position, a number of non-aligned countries like China, India, Iran and Pakistan argue that only declared facilities should be subjected to visits.[49]

Three different types of visits – randomly-selected transparency visits, voluntary assistance visits, and declaration clarification visits – were eventually included in the text of the Compliance Protocol, which stipulates in Article 6, paragraph 5 that the total number of all visits shall not exceed 120 per calendar year. Randomly-selected transparency visits are to increase confidence in the consistency of declarations with activities going on at the declared facilities, and to increase the transparency of declared facilities. Voluntary assistance visits allow a state party to invite the inspectorate of the BWC organization to conduct a visit on its territory in order to obtain technical assistance and information related to the implementation of the obligations under the Protocol or otherwise. In contrast, declaration clarification visits or procedures serve to clarify a potential omission of a facility in a states party's annual declaration.

Of the overall amount of visits, between 60 and 90 will be randomly-selected transparency visits, between six and 30 will be voluntary assistance visits, and a maximum of 54 visits can be devoted to declaration clarification procedures. In order to ensure the equitable distribution of visits among states parties to the Protocol and to avoid placing too high a burden on individual states or facilities, several precautions have been taken, placing upper and lower limits on the number of visits to states and facilities. Especially for randomly-selected transparency visits and for clarification visits, the Protocol contains detailed guidelines for the initiation, conduct and reporting on these types of visits.

The three types of visits put emphasis on different aspects of the follow-up procedures to declarations to be submitted by states parties. They also present differing degrees of intrusiveness with voluntary assistance visits – to be initiated by the visited state party itself – at the low end and declaration clarification procedures – to check on facilities that should have been declared but were not – at the upper end of impacting on a state's sovereignty and information security. They thereby offer a set of tools for bringing to life the transparency norm, allowing states parties to demonstrate their compliance with the non-acquisition norm, and presenting an additional procedure for putting into effect the cooperation norm – all of which are absent from the current BWC prohibition regime.

4.3.3. Investigations

The purpose of investigations in general terms is to provide the regime with the means to check on suspected non-compliant behaviour by states parties to the Protocol. Investigations are not a tool for ascertaining treaty compliant behaviour. Compared to the usage of the term 'inspections' in the CWC context, investigations in the context of the BWC Compliance Protocol only refer to the equivalent of CWC challenge inspections. The concept of routine inspections as found in the CWC can be compared with the visits-concept, but even here some delegations in the AHG – like the US and some NAM countries – were throughout the negotiations very hesitant to agree on anything close to routine on-site measures, whatever the terminology attached to them might be.

Another fundamental issue, which AHG members were unable to agree upon, relates to the initiation of an investigation and the related question of how an investigation that might be frivolous can be stopped. Basically, two different approaches could be identified. According to the first one, the so called 'red-light' approach, an investigation request will go forward unless a majority – which would have to be defined depending on the circumstances of the request – of the executive council of the future organization decides otherwise. This position was supported by states – like the UK – who wish to see a low threshold established for such requests and expect an effective compliance regime to result from such a procedure. The 'green-light' approach, in contrast, would allow for an inspection to proceed only if a majority of EC members approves it. This is championed by those states in the AHG who want to raise the threshold for triggering investigations and thereby expect to prevent frivolous challenges or minimize the risk to national security assets or of industrial espionage.[50]

The Protocol in Article 9 distinguishes between two types of investigations: field and facility investigations.[51] While the latter one focuses on 'the perimeter around a particular facility at which there is a substantive basis' for a non-compliance concern related to Article 1 of the BWC, field investigations are concerned with larger geographic areas, where either similar cause for a non-compliance concern exists or a disease outbreak might be directly related to activities prohibited by the Convention.[52]

Before initiating an investigation states parties are required to consult among themselves with the aim of resolving any matter related to a non-compliance concern – the details of the available consultation, clarification and cooperation procedures being laid out in Article 8 of the Protocol. Only when these measures cannot dispel a non-compliance concern shall a state party initiate an investigation. Depending on the type of investigation requested, the kind of non-compliance concern forming the basis for the request, and the location at which the investigation is to be conducted, different decision-making procedures are to be applied.

In all of the above five scenarios the receiving state party, that is the state on whose territory the investigation is taking place, has the right to determine the nature and extent of access that is being granted to the investigation team. In addition to this 'managed access' procedure, Article 9 contains detailed provisions for moving from one type of investigation to the other, spells out measures to guard against abuse of investigations, provides detailed timelines for the conduct of investigations, and outlines procedures for reporting the findings of an investigation and for follow-up activities.

4.4. Cooperation in peaceful uses and export controls

According to critics of export controls, in the BWC context such measures are in direct contravention to Article X of the BWC, which in its Paragraph 2 establishes the cooperation norm and states that 'this Convention shall be implemented in a manner designed to avoid hampering the economic or technological developments of States Parties to the Convention or international cooperation in the field of bacteriological (biological) activities'. Proponents of export controls, in contrast, point out that these measures are an expression of their implementing the non-transfer norm contained in the Convention's Article III, according to which states parties are under the obligation not to transfer to any recipient whatsoever – directly or indirectly – any of the agents, toxins, weapons, equipments or means of delivery specified in Article I of the Convention.

Measures to enhance the implementation of Article III were discussed in some detail during the March 1997 session of the Ad-Hoc Group, when India introduced a Working Paper setting out a number of guidelines to strengthen the non-transfer norm.[53] Were the Indian proposal be put into effect, this amounted to nothing less than abrogating the present export control practice, including a substantial transfer of national sovereignty to a future BWC organization, which would be empowered to decide on all BW-related transfers to states not party to the Protocol. By implication, transfers among members to the Protocol are free on principle. This would strip individual member states of their decision-making prerogative and would make the future BWC organization the central decision-making organ. Following from that, if national export controls became obsolete, so would multilaterally agreed-upon controls, i.e. those of the Australia Group.

Austria and New Zealand, in contrast, pursued a more moderate approach to strengthening Article III of the BWC. In a Working Paper the two Australia Group members advocated to include in the future protocol an obligation for each state party to declare annually 'the national legal measures it has adopted in order to implement Article III of the BWC' and to 'report to the Organization on an annual basis on its administrative and other related national implementation measures with regard to Article III of the BWC to ensure that transfers of agents, toxins, and equipment are only authorized in compliance with the provisions of the Convention'.[54] If such measures were implemented, not only would the decision-making power continue to reside with the states parties, but the declarations would also be limited to national measures. Multilateral measures such as the Australia Group would not be touched and consequently would not become more transparent to states outside this group.

The second major bone of contention in the deliberations of the AHG on Article 7 was related to a NAM-proposal to establish a Cooperation Committee within the organization appointed to oversee the implementation of the Protocol.[55] The Committee's task would have been to coordinate and promote effective and full implementation of Article X of the Convention and Article 7 of the

Protocol. This had gradually come to be accepted by all negotiators in the AHG, even the US delegation, which was waiting to see the details of what the proposal would entail before stating its position. Eventually, during the 18th negotiating session in late 1999 the US delegation agreed in principle to the establishment of such a Committee.[56] Still, disagreement remained as to the structure and precise roles and functions of the Cooperation Committee, and the powers allocated to it. It can be assumed that this institutionalized mechanism for cooperation in the peaceful uses of the biosciences was judged by AHG negotiators to be one of the bargaining chips for the end-game of negotiations, in order to be traded against concessions in the field of export controls or compliance mechanisms, areas in which some of the Cooperation Committee's proponents were reluctant to agree on wide-ranging and intrusive measures.

In Article 7 of the Composite Text states parties are tasked to review, amend or establish 'any legislation, regulatory or administrative provisions to regulate the transfer of agents, toxins, equipment and technologies relevant to Article III of the Convention'.[57] The Protocol further requires states parties to establish transfer guidelines with the aim of ensuring that transfers of dual-use items will be used for permitted purposes only. Accordingly, the supplying state shall require from its recipient an end-use declaration, a no-retransfer pledge (without prior consent of the originating state party) and information from the requesting state on its national laws and regulations pertaining to the items in question. What is more, states parties are under the obligation to report annually on the transfers they have undertaken with respect to four categories of dual-use items: (1) fermenters or bioreactors with a volume of at least 100 litres, (2) aerosol chambers, (3) equipment for use in aerobiology studies that can generate particles of 20 or less microns in diameter, and (4) analytical equipment for determining the size of aerosol particles of 20 or less microns in diameter.[58] In a separate section on consultations, the Protocol clearly spells out that states parties are allowed to 'consult among themselves on the implementation of the provisions' thereby implicitly acknowledging the future existence of groups like for example the Australia Group.

Article 14 of the Compliance Protocol contains the provisions to strengthen Article X of the BWC on the peaceful uses of the biosciences. Here the section on 'Institutional Mechanisms for

International Co-operation and Protocol Implementation Assistance'[59] contains a useful compromise wording concerning the differing approaches to the functions and powers of the Cooperation Committee. Accordingly, this Committee is designed to 'consult on, monitor and review activities fostering international co-operation and assistance'. Its output is limited to reports and its tasks do not include decision-making of any kind. Likewise, the review of concerns raised by a state party that it has been deprived of benefits stemming from Article X of the BWC is envisaged to be reviewed by the Executive Council of a future BWC-organization.

Both the compromise solution adopted in the Protocol with respect to the non-transfer norm and the cooperation norm add not only a set of rules and procedures for putting into effect the more abstract norms into everyday state practice. What is more, this regulatory underpinning to the two norms would have increased transparency of states parties' behaviour while at the same time safeguarding their national decision-making prerogatives.

4.5. The collapse of the AHG negotiations

The Fourth BWC Review Conference in 1996 clearly expressed its expectation that the AHG negotiations on a Compliance Protocol would be concluded and a Special Session on the Protocol be held before the Fifth Review Conference in 2001.[60] In the intervening five years between the two Conferences many political statements provided additional support to the realization of a Protocol in this time frame. Thus, it did not come as a surprise when the Chairman of the AHG, Ambassador Toth, set out in early 2001 to present his version of a compromise text. After extensive consultations this chairman's draft was finalized and presented to all participating delegations in March 2001.

The July 2001 session of the AHG was scheduled to have a debate of the compromise text submitted by the AHG chairman. While not many delegations were satisfied with all aspects of the draft protocol presented by Ambassador Toth, the US delegate, Ambassador Donald Mahley, concluded in his remarks:

> that the Chairman's text was not an adequate basis for completing the Protocol, that it could not be made an adequate base for further negotiation and, furthermore, that the whole conceptual

> framework on which the negotiations had been conducted, would have to be changed.[61]

The US statement, which effectively dealt the death-blow to the AHG negotiations, marked the culmination of a difficult history of successive US administrations with efforts to strengthen the BWC. This history had been characterized by difficulties to formulate a coherent government policy on the BWC protocol in the first place, continuous intra-agency quarrels over the course US policy should take, and, resulting from this, numerous US demands vis-à-vis the AHG, which *de facto* lead to a weakening of several provisions of the Protocol. In an almost ironic twist of argumentation one of the basic criticisms levelled against the Protocol was that it was too weak and thus would not provide for effective verification. In addition, the US delegate argued, the Protocol 'would threaten national security and commercial proprietary information; and it would threaten the dual-use export control regime of the Australia Group'.[62] All of these criticisms have been subjected to a critical analysis.[63]

5. The BW prohibition regime since the 5th BWC Review Conference

5.1. The split 5th Review Conference

When the 5th BWC Review Conference convened in November 2001 the United States decided in a rather undiplomatic move to accuse a number of states, among them BWC states parties Iran, Iraq and Libya, of clandestinely procuring BW. As one commentator observed:

> The US decision to 'name names', an unorthodox diplomatic proceeding, took many delegates and observers by surprise. Overall, there was a feeling that the accusations have only compounded the already tense and bitter atmosphere left over from the derailed final session of the AHG in July/August.[64]

Yet, despite this difficult start, the diplomatic machinery of the Review Conference worked well over the following two weeks and good progress was made on a consensual final declaration.[65] On the very last day of the Conference the US came forward with a proposal to terminate the AHG for good. The content of this proposal ran counter to the tacit understanding of not touching the topic of the AHG in order to avoid a breakdown of the review process as well. What is more, the US obviously did not inform in advance any of its allies about the content or timing of the proposal. Not surprisingly, this created the impression that the US delegation deliberately attempted to wreck the Review Conference. The only way to prevent a diplomatic disaster was to adjourn the Conference and to decide to reconvene one year later in November 2002.[66]

Despite rumours that the US government planned to have the Conference only reconvene for one day in order to decide to meet again in 2006 for the next Review Conference, the Conference did last somewhat longer and achieved more than just agreeing on the date for the next meeting. However, it fell far short of deciding on a final document, which would have represented the consensual interpretations of states parties on the BWC and its implementation. What was eventually achieved, falls somewhere between these two poles on a spectrum of possible outcomes.[67] The chairman of the Review Conference presented (upon US insistence) a non-negotiable draft decision to the conference which leaves the AHG mandate untouched and establishes a 'new process' to guide activities for the years 2003 to 2005. This 'take-it-or-leave-it'-proposal was debated for several days and eventually adopted on 14 November 2002. In it states parties agree:

(a) To hold three annual meetings of the States Parties of one week duration each year commencing in 2003 until the Sixth Review Conference, to be held not later than the end of 2006, to discuss, and promote common understanding and effective action on:

  i. the adoption of necessary national measures to implement the prohibitions set forth in the Convention, including the enactment of penal legislation;
  ii. national mechanisms to establish and maintain the security and oversight of pathogenic microorganisms and toxins;
  iii. enhancing international capabilities for responding to, investigating and mitigating the effects of cases of alleged use of biological or toxin weapons or suspicious outbreaks of disease;
  iv. strengthening and broadening national and international institutional efforts and existing mechanisms for the surveillance,

detection, diagnosis and combating of infectious diseases affecting humans, animals, and plants;

v. the content, promulgation, and adoption of codes of conduct for scientists.[68]

The first two of these topics was dealt with in 2003, items iii and iv on the list were covered in 2004 and the remaining one – codes of conduct – has been addressed in 2005. Each meeting of states parties is preceded by a two-week expert meeting, the latter of which are allowed to produce 'factual reports describing their work'. The Sixth Review Conference, to be held no later than the end of 2006, has been tasked to 'consider the work of these meetings and consider any further action'. This programme of work clearly is a far cry from the comprehensive approach of the AHG to reach agreement on a legally binding Protocol. Yet, at least it keeps the diplomatic machinery that is concerned with strengthening the BWC going and allows some topics to be addressed by states parties. The following section will briefly look at the results this 'new process' has yielded so far.

5.2. Attempted strengthening of the regime II? – The 'new process' in action

The 2003 meetings of experts and states parties were held in August and November of that year, respectively. The topics to be discussed were 'the adoption of necessary national measures to implement the prohibitions set forth in the Convention, including the enactment of penal legislation' and 'national mechanisms to establish and maintain oversight of pathogenic micro-organisms and toxins'. The most positive aspect of this first round of the new process was that during the experts meeting many states parties submitted detailed information as regards their national experience in the above mentioned two areas. This has clearly increased transparency and marks a welcome deviation from the pattern of widespread ignorance established in relation to the politically binding CBMs that were agreed upon in 1986 and 1991. However, as one commentator pointed out, this 'mountain of paper' was delivered some 23 years late, as already the First BWC Review Conference in 1980 had issued in its final document a call to submit such information.[69] While the experts meeting limited itself to a technical discussion of the issues involved, there was hope that during the 'annual meeting a political assessment of its outcome and more detailed exploration of how to take things forward' would be conducted.[70] As a matter of fact, during the 2003 meeting of states parties:

> several countries (including Germany, India, New Zealand, Pakistan, South Africa and Sweden) argued for distilling the voluminous data exchanged at the experts' meeting into a set of voluntary guidelines for penal legislation and biosecurity regulations, that could be incorporated into the final document, thereby ensuring greater uniformity and consistency in how member states implement the BWC.[71]

However, as all decisions in this new process have to be made by consensus, a few states parties, which objected to the identification of such 'best practices', were able to prevent this from happening. As a result, what has been described as the 'substantive' part of the political statement merely urged member states to do what their ratification of the BWC under Article IV of the Convention requires them to do anyway – enact the BWC's prohibitions on the domestic level. Although a few pointers are contained in the statement of the states parties meeting, there is no concrete guidance offered or harmonization sought. Eventually, those states parties who were rejecting any form of obligations resulting from the new process succeeded in their efforts.

That this pattern would repeat itself in 2004 appeared likely already during the opening of the experts meeting in July that year, when Cuba for example insisted in its statement during the general debate that no recommendations should be formulated based on the experience of a few countries only. Likewise, the delegate of the Republic of Korea pointed out that the meeting was conceptualized as a forum to exchange ideas, not as a drafting exercise. India, for its part emphasized that the mandate of the new process foresaw to promote, but not to reach a common understanding, and that the latter would require some form of negotiation, which was not part of the mandate the BWC states parties had given themselves for the intersessional period leading up to the 6th BWC Review Conference in 2006. Despite these attempts to limit what would be achievable during the second year of the intersessional process, some commentators saw a more constructive approach prevailing in 2004 than the year before:

> The atmosphere at the 2004 Meeting of States Parties was more positive than a year ago. It was particularly noticeable in the statements in the General Debate, as well as in some of the NGO statements and activities, that attention is increasingly being given to the Sixth Review Conference in 2006.[72]

It remains to be seen however, whether the positive atmosphere – which was again displayed during the 2005 meeting of experts – can be carried over into the 6th Review Conference in 2006 and whether any meaningful strengthening of the BW prohibition regime will result that adequately takes into account the revolution in the life sciences. Paying merely lip-service that the scope of the BWC covers all S&T developments will not suffice.

5.3. Parallel controls of science and technology

During the same period in which the negotiations of the Ad-Hoc Group, which negotiated a legally binding protocol to the BWC, came to an end, and the Fifth BWC Review Conference struggled to produce any constructive outcome at all, several developments in the life sciences took place which many observers saw as opening wide the door for potential misuse. The 'contentious research' in question involved:[73]

- unintentionally potentiating the virulence of the mousepox virus through inserting an IL-4 gene into the mousepox genome;
- synthesis of the poliovirus genome from 'chemically synthesized oligonucleotides that were linked together and then transfected into cells', thereby creating an infectious virus from scratch;[74]
- transfer of a virulence factor of *variola major* (which causes smallpox) into the *vaccinia* virus, which is of much lower virulence and usually used for vaccinations against smallpox.

Concerns expressed over these experiments in the media and the political system (mostly in the United States) led the National Academies of Science to establish a committee to investigate ways to prevent S&T advances from being misused for hostile purposes. The so-called Fink Committee issued a set of recommendations to address the new environment in which the life sciences are operating and to prevent scientific advances from being misused by states or terrorist groups in BW programmes, while at the same time 'enabling legitimate research to be conducted'.[75] The Fink Committee's seven recommendations are:

> ... that national and international professional societies create programs to educate scientists about the nature of the dual-use dilemma in biotechnology and their responsibilities to mitigate its risks. ...
>
> ... that the Department of Health and Human Services (DHHS) augment the already established system for review of experiments involving recombinant DNA conducted by the National Institutes of Health to create a review system for seven classes of experiments (the Experiments of Concern) involving microbial agents that raise concerns about their potential for misuse. ...
>
> ... relying on self-governance by scientists and scientific journals to review publications for their potential national security risks. ...
>
> ... that the Department of Health and Human Services creates a National Science Advisory Board for Biodefense (NSABB) to provide advice, guidance, and leadership for the system of review and oversight we are proposing. ...
>
> ... that the federal government rely on the implementation of current legislation and regulation, with periodic review by the NSABB, to provide protection of biological materials and supervision of personnel working with these materials. ...
>
> ... that the national security and law enforcement communities develop new channels of sustained communication with the life sciences community about how to mitigate the risks of bioterrorism. ...
>
> ... that the international policymaking and scientific communities create an International Forum on Biosecurity to develop and promote harmonized national, regional, and international measures that will provide a counterpart to the system we recommend for the United States.[76]

The NSABB has in the meantime been established in the office of the director of the National Institutes of Health.[77] The functions of the

NSABB are to 'advise the Secretary of HHS, the Director of NIH, and the heads of all federal departments and agencies that conduct or support life science research. The NSABB will advise on and recommend specific strategies for the efficient and effective oversight of federally conducted or supported dual-use biological research, taking into consideration both national security concerns and the needs of the research community.'[78] The NSABB will be composed of a maximum of 25 voting members whose areas of expertise will cover *inter alia* genomics, bacteriology, virology, laboratory biosafety and biosecurity, public health, pharmaceutical production, bioethics, national security, intelligence and law enforcement. In addition, more than a dozen government departments and agencies will be ex officio members of the board.[79]

With a view to the second recommendation concerning restrictions in the publication of problematic research a number of journal editors had imposed restrictions on themselves already before the publication of the Fink Committee's report: in January 2003 a group of 32 journal editors agreed on guidelines related to 'Scientific Publication and Security'. After first being published in *Science*, the statement also appeared in February in the *Proceedings of the National Academy of Sciences* and in *Nature*.[80] The authors of the statement:

> recognize that the prospect of bioterrorism has raised legitimate concerns about the potential abuse of published research, but also recognize that research in the very same fields will be critical to society in meeting the challenges of defense. ... We recognize that on occasion an editor may conclude that the potential harm of publication outweighs the potential societal benefits. Under such circumstances, the paper should be modified, or not be published.[81]

6. Summary and conclusion

In comparison to the CW prohibition regime analysed in the previous chapter, the BW prohibition regime obviously is much less developed and thus even less equipped to deal with the scientific and technological challenges that lie ahead. As previous research has shown, during the 20th century offensive military BW programmes utilized the state of the art knowledge in the life sciences available at that point in time. The current biotechnology revolution necessitates shoring up the regime in order to be able to address the potential for misuse of 21st century life sciences.

Statements by successive review conferences to reconfirm that S&T developments are still covered by the scope of BWC Article I and thus fall under the prohibitions contained in the Convention are a laudable and necessary effort to uphold the norms against the misuse of the life sciences. However, the formulation of ever lengthier interpretations of the BWC's scope has long been recognized as being insufficient to tackle the regime's basic problems; hence the creation of the Ad-Hoc Group of states parties to negotiate a legally binding protocol to the BWC. From the point of view of preventing the misuse of advances in S&T the protocol would have provided two major benefits: first, it would have increased transparency with respect to crucial research in the life sciences, and secondly, it would have created an organizational structure in which to address S&T related issues in a more systematic and efficient way. With the collapse of AHG negotiations, however, even a limited infrastructure such as a small secretariat or the Scientific Advisory Board in the CWC context is not available in the BW prohibition regime.

In the so-called 'new process' only a few selected issues are addressed. Thus it lacks the comprehensive approach of the AHG process. What is more, as the first round of this new process in 2003 has demonstrated, even on such a limited scale several states parties insisted on adopting an approach which would not result in any obligations for them. Resulting from this, the already existing wide gap between science and technology in the life sciences racing ahead and the prohibition regime being stuck in the standards of 1960s type arms control measures is bound to widen.

This is not going to be changed by the parallel controls of scientific and technological development being instituted in some states, most notably the US, as these attempts are uncoordinated internationally and as decoupled from overall regime development as is the new process. Even if an attempt should be made by the US government to externalize its current policy, which aims at providing biosecurity on the national level, this is both bound to run into opposition from several BWC states parties, which will object to the idea of having to accept yet another US *fait accompli* and it is going to fall short of the required overall regime strengthening – given the revolution of the life sciences that lies still ahead of us. It is some of the dimensions of this revolutionary development which we will turn to in the next three chapters, analysing developments in three selected areas of the life sciences, which we believe will be of fundamental importance in this regard: immunology, neurology and neuro-endocrino-immunology.

4
Defences Under Attack: the Potential Misuse of Immunology

1. Introduction

The immune system plays a crucial role in protecting against infectious diseases. This is clearly demonstrated in the case of individuals with genetic defects in certain immune mechanisms, which frequently result in a devastating outcome, despite the use of antibiotics or other chemotherapeutic agents. Furthermore, some microorganisms can escape immune defences by using strategies that subvert immune mechanisms. These strategies represent factors that contribute critically to the pathogenicity of the microorganism, or its ability to cause disease. Indeed, the pathogenicity of a microorganism can only rightly be defined within the scope of its interaction with the immune system.

In this age of rapid biomedical and biotechnological advances, far-reaching manipulations of microorganisms are now possible that can change their properties drastically. Experiments to manipulate microorganisms are being carried out daily, with mostly peaceful aims in mind, such as the elucidation of the pathogenic mechanisms of an infectious agent, which could in turn point the way to the development of better prophylactic and therapeutic measures to counter infections more successfully.

However, it has become evident that these experiments can lead to the creation of particularly dangerous microorganisms that can evade the immune responses in devastating ways. Some examples will be offered to illustrate this point, which will then be handled more thoroughly in a later section. A prime example is the inadvertent creation of a killer mousepox virus by researchers trying to develop a contraceptive vaccine to control the rodent population in Australia.[1]

Particularly disturbing is the fact that another scientist, Mark Buller at Saint Louis University, has picked up on these experiments and carried them one step further by increasing the lethality of the mousepox virus and by planning to carry out similar manipulations in the cowpox virus.[2]

Another example concerns the potentiation of a virulence factor of vaccinia virus, the strain used for smallpox vaccinations.[3] In this case a property (factor) of the virus was engineered to more effectively eliminate an essential component of the innate immune system, which acts as the first line of defence against an infection.

In a third example, Russian researchers transferred toxin genes from a soil bacterium to *Bacillus anthracis*, the causative agent of anthrax.[4] The pathogenicity of the engineered bacterium was not increased by this manipulation per se, but hamsters that had been vaccinated against anthrax were not protected after challenge with the engineered strain. In some way which was never clarified, the engineered agent was able to evade immune defences normally afforded by vaccination.

In addition to microorganisms attacking the immune system, toxins are also of particular concern. Prominent among these is the bacterial product *Staphylococcus* enterotoxin B (SEB), which is a potential biological weapon that also falls into the category of a potential chemical weapon. This toxin was, for example, on the US list of favoured anti-personnel agents as early as 1949[5] and was apparently weaponized by the US Army prior to the Biological Weapons Convention.[6]

Up to now, the focus has been mainly on concerns about the possibilities of manipulating the properties of microorganisms to make them more robust and more pathogenic. It becomes evident from the examples cited above that the real target in all these cases is the immune system, and how vulnerable it is to evasion mechanisms, which naturally potentiate the pathogenicity of the infective agents. This represents a change of focus from the microorganism to systems biology and how it might be misused. It is also evident that with the rapid expansion of research activities in the areas of molecular biology and biotechnology advances occur at an exponential rate, along with increasing capabilities for misuse.

The situation is accentuated by the fact that the National Institute of Allergy and Infectious Diseases (NIAID) of the US National

Institutes of Health (NIH) has expanded its programme significantly in order to attract scientists to the area of biodefence research. 'In FY 2002 and FY 2003, NIAID developed more than 50 initiatives to stimulate biodefense research; approximately 75% of these are new initiatives and 25% are significant expansions of existing contracts.'[7] Within this programme, immunology as it relates to biodefence is given special attention. NIAID reported that it awarded a multicomponent grant to create an 'encyclopedia' of innate immunity, a comprehensive and detailed picture of the type of immunity that represents the essential first line of defence against infectious diseases. The stated goal of this undertaking is to gain knowledge that could lead to the development of treatments for infectious diseases. At the same time, however, this information could provide a blueprint for malign attack of the immune system.

In order to appreciate the dilemma of dual use and the possibilities of misuse in this area, a brief description of scientific and technological aspects underlying research activities in this field, including the elements of the innate and the acquired immune systems will be given. Also, mechanisms of immune evasion used by some microorganisms will be outlined. With this background, examples of research in which microorganisms have been created that evade immune defences will be presented along with an analysis of the dual use aspects involved. Finally, a look to possible future threats involving the vulnerability of the immune system will be offered.

2. Scientific and technological background

2.1. Mammalian immune system

The hallmark of the immune system is its ability to respond to an invasion of the body by microorganisms or toxic components in ways that afford protection against the detrimental effects that could occur. The responses of the immune system include both specific (adaptive immune system) and non-specific (innate immune system) components. These react in different ways to antigens, which are substances foreign to the host. For example, antigens can be protein or polysaccharide components of microorganisms, but also toxic substances produced by them. Components of innate and adaptive immunity are listed in Table 4.1.

Table 4.1: Features of innate and adaptive (specific) immunity

Feature	Innate immunity	Adaptive immunity
Characteristics		
Specificity for microorganisms	Relatively low (PAMPs)[a]	High (specific antigens)
Diversity	Limited	Large
Specialization	Relatively stereotypic	Highly specialized
Memory	No	Yes
Components		
Physical and chemical barriers	Skin, mucosal epithelia; anti-microbial chemicals e.g. defensins	Cutaneous and mucosal immune systems; secreted antibodies
Blood proteins	Complement	Antibodies
Cells	Phagocytes (macrophages, neutrophils), Natural killer cells	Lymphocytes (B cells that produce antibodies, T cells that carry out cell-mediated reactions

Source: From A. K. Abbas, A. H. Lichtman and J. S. Pober, *Cellular and Molecular Immunology*, 3rd edition (Philadelphia: W. B. Saunders Company, 1997).
[a] PAMPs: pathogen-associated molecular patterns.

The innate immune system includes components that are present and ready for action even before an antigen challenge is encountered. These are cellular and molecular components that are less specific than those of the adaptive system. Macrophages, for example, are phagocytic cells that represent a prominent cellular component of innate immunity. These cells do not recognize antigens in a specific manner but react to classes of antigenic substances from microorganisms called pathogen-associated molecular patterns or PAMPs. A simple analogy using car models and a specific manufacturer can be used to illustrate. All models of Volkswagen cars carry an identical VW emblem. A PAMP is like the emblem, which is present on all different models of VW vehicles. Any vehicle carrying this emblem would be recognized as manufactured by Volkswagen. However, this emblem provides no information as to the particular model of vehicle. This is very similar to the way in which the innate immune system recognizes many different microorganisms carrying a particular PAMP as a class of microorganism, but it is not able to identify the particular microorganism. The adaptive immune system, on the other hand, is able to distinguish one particular microorganism from another by recognizing other, more specific or distinctive, features of the model.

Several components of the innate immune system must be activated by agonists such as PAMPs, but this activation can occur within minutes or hours rather than days. Therefore, innate reponses are quicker, but the immunity they afford may not be as effective over as long a period of time as adaptive immunity. Nevertheless, the innate immune system represents the all-important first line of defence against pathogens and is absolutely essential for keeping an infection in check before adaptive immunity can be induced. If innate immunity is malignly attacked, the battle against infections is lost from the start.

In general, two main types of adaptive or aquired immune responses may occur: humoral and cell-mediated,[8] involving the action of B and T lymphocytes (Table 4.2). These are the so-called immunocompetent cells of the immune system, because they are able to react to an antigen challenge with a high degree of specificity. Activation of lymphocytes occurs through engagement of receptors to specific antigens on the cell surface. In the case of B cells, these receptors are membrane-bound antibodies. The antigen receptors of T cells are called the T cell receptor (TCR). T cells are further subdivided into T helper cells (Th), which carry the identifying CD4 molecule on the surface and cytotoxic T cells (CTL or Tc), which carry the CD8 identifying molecule.

Whereas antibodies can recognize and bind antigen alone, the T cell receptors can not; the antigens that T cells interact with specifically have been processed into antigen fragments and then bound to molecules called major histocompatibility complex molecules or MHC molecules, which are exported to the surface of the antigen processing cells. In this respect, it is said that the antigen fragments are presented to the T cells by the MHC molecules on the surface of the antigen presenting cell. T cells therefore have a double recognition requirement: foreign antigen and self MHC molecules. There are two classes of MHC molecules (I and II) involved in antigen presentation. There are mainly two types of T cells, helper (Th) and cytotoxic (Tc). Just as the name implies, Th cells assist other cells (such as B cells, Tc cells and macrophages) to carry out their effector functions. Th cells accomplish this by providing activating signals through direct interaction with the other cells and by secreting substances (cytokines) that have various effects on the interacting cells. Th cells recognize antigen fragments

Table 4.2: Features of the adaptive immune system

		T lymphocytes		
Feature	**B lymphocytes**	**Th1**	**Th2**	**Tc**
Function	Antibody production	Help for T cells and macrophages (cell-mediated responses)	Help for B cells (antibody-mediated responses)	Cytotoxic action against virus-infected cells and tumour cells
Receptor for antigen	Membrane-bound antibodies	T cell Receptor (TCR)	TCR	TCR
Subset surface molecules	Different antibody classes	CD4	CD4	CD8
Antigen recognition by receptors	Free antigens (proteins, polysaccharides)	Protein fragments bound to MHC II molecules	Protein fragments bound to MHC II molecules	Protein fragments bound to MHC I molecules
Soluble substances produced	Antibodies	Cytokines e.g. IL-2, IFNγ	Cytokines e.g. IL-4, IL-10	Perforin, Granzymes
Co-stimulating molecules	B7, CD40	CD28, CD40L	CD28, CD40L	CD28, CD40L

Abbreviations: Th, helper T cells; Tc, cytotoxic T cells; MHC, major histocompatibility complex; IL, interleukin; IFN, interferon; CD, a designation for antigenic molecules on the surface of cells.

presented by MHC II molecules, which are found on only a few types of cells in the body, mainly B cells, macrophages and macrophage-like dendritic cells. Tc cells on the other hand recognize cells that have been parasitized, by sensing antigen fragments bound to MHC I molecules that are found on all cells of the body with a nucleus. This means, practically speaking, that all cells can present antigen to Tc cells in complex with MHC I molecules. Tc cells can attack, damage and kill cells that have become invaded by pathogens (particularly viruses, but also some bacteria, fungi and protozoa), thus eliminating the production sites before these pathogens can multiply appreciably.

Antigenic signals are transduced from the receptors over signal cascades that are activated in the inner part of the cell, leading in the end to the expression of genes controlling the biosynthesis of products of the cell. This activation of lymphocytes to effector cells usually takes five to six days, resulting in the production of antibodies by the B lymphocytes and other effector molecules by the T lymphocytes. In the course of activation, so-called memory cells of both B and T lymphocyte types are developed, which can respond more quickly to antigen during a secondary or later challenge. Thus, adaptive immunity affords a high degree of protection, but it takes time to be induced.

Macrophages occupy a central position in the immune system, being active both in innate and adaptive immune responses. In innate immunity, macrophages are activated through engagement of receptors on the cell surface by substances called agonists. Most prominent among receptors on the macrophage surface are the Toll-like receptors (TLRs). The TLRs derive their name from the similarity with the transmembrane receptor protein Toll in the fruit fly *Drosophila*, which is involved in development and in protecting flies against fungal infections. This has been termed 'an ancient system of host defense'.[9]

Up to now, several different TLRs in humans have been described. These molecules contain a characteristic leucine-rich extracellular domain (LLR), which recognizes conserved structures of the micro-organisms called pathogen-associated molecular patterns (PAMPs), referred to above under the discussion of innate immunity. This initial reaction between TLRs and PAMPs leads through a signalling cascade to the activation of genes that control the production of the proinflammatory cytokines.[10]

Macrophages produce type I interferons (α and β), which are essential for a successful defence against many viral infections. They are also potent producers of proinflammatory cytokines including interleukin 1 beta (IL-1β), IL-6 and tumour necrosis factor alpha (TNFα), which mediate reactions designed to fight infections. When these cytokines are produced in moderate amounts, they induce mild inflammation reactions and contribute greatly to defence mechanisms directed against pathogens and to the healing process in general. If they are produced in particularly large amounts or continually during chronic illnesses, this can lead to various disorders such as coronary insufficiency, thrombus formation, hypoglycaemia, and in some cases even to shock and death.[11] This makes these activities particularly vulnerable to malign modulation such as by targeting the TLRs to induce hyper-responses, or by inhibiting key components in signalling cascades that would upset the balance. It is significant in this respect that IL-1 was reported to be effective in aerosol form in pulmonary absorption studies carried out by the US Army under its medical research programme.[12]

Macrophages bridge innate and adaptive immunity. After they have devoured foreign antigens or microbes as part of their role in innate immunity, they assist B cells and T cells in adaptive responses by producing cytokines that regulate lymphocyte function or by presenting antigens bound on MHC molecules so that these antigens can be recognized by T cells. Furthermore, they increase other substances (called co-stimulatory molecules) on their cell surface that can generally enhance their interaction with T cells.

2.2. Innate immunity of plants

As biological warfare can be directed against plants as well as humans and animals, a brief description of immune mechanisms in plants and their possible misuse will follow.

Plants also exhibit a type of innate immunity, revealed by their resistance to certain pathogens.[13] Essentially two kinds of reactions are recognized. One is cultivar-specific, that is, the reaction occurs only with particular plant varieties (cultivars) and some kinds of pathogen species. This is genetically determined and involves complementary pairs of pathogen-encoded avirulence genes (*AVR*) and plant-encoded resistance (*R*) genes. The interaction of AVR proteins with plant R proteins elicits plant defence reactions. The other kind of reaction involves a large variety of microbe-associated products which can trigger defence responses in many plant species in a non-cultivar-specific manner. These products include an array of molecular components such as oligochitins, glucans, peptides or lipopolysaccharides, resembling the PAMPs described above for mammalian systems. The vast majority of plant R proteins that have been characterized resemble modular structures of the LRR-containing Toll-like receptors (except that the intracellular domains are different from those of the TLRs), or the more recently discovered intracellular nucleotide-binding oligomerization domain (Nod)-LRR proteins also implicated in PAMP recognition in humans.[14]

A number of physiological changes are known to occur in response to pathogen attack on plants, including the production of reactive oxygen species (ROS), release of secondary signal molecules such as nitric oxide (NO) and the synthesis of antimicrobial products including phytoalexins and pathogenesis-related (PR) proteins.[15] The PR proteins include substances which are known to have anti-fungal and anti-bacterial properties.

One of the most prominent responses is the hypersensitive reaction (HR), which involves the production of H_20_2 and is characterized by rapid and localized cell death at the infection site, serving to limit the spread of the infection. A heightened, systemic resistance can be observed after secondary attack by a broad range of plant pathogens, and this type of immunity is called systemic-acquired resistance (SAR).[16] The main systemic signals include salicylic acid, jasmonate and ethylene, which are produced in response to wounding and insect attack.[17] H_20_2 is, however, the most important ROS involved in downstream signalling, leading to the activation of signalling cascades in *Arabidopsis* as well as activation of genes controlling the production of proteins involved in HR.

Similar to the macrophages discussed above, plants may be attacked through their innate immune systems, for example by targeting either the receptors of signalling cascades, or by inhibiting or producing an over-reaction in a signalling cascade with the use of inhibitors of key components in that cascade.

2.3 Immune evasion by microorganisms

In order for a microorganism to be pathogenic, it must have a mechanism that permits it to evade immune defences. There is a great deal of interest in studying these processes with the aim of developing means of countering evasion strategies, which would permit, for example, the development of vaccines that defeat the evasion tactics of antigenic variation used by microorganisms. At the same time, exploitation of evasion strategies with malign intent should be of particular concern. Some evasion strategies are described below.

2.3.1. Antigenic variation

Some microorganisms frequently mutate or vary their antigenic composition so that they can no longer be recognized by the antigen receptors of immune system cells. With regard to particular antigens, some microorganisms exhibit a much higher mutation rate than is normal. This is encountered, for example, in connection with the flu virus, the AIDS virus or the causative agent of Lyme disease, *Borrelia burgdorferi*. This is one reason these infectious diseases are resistant to vaccination. In addition, some microorganisms are subject to mutation due to pressures exerted by the immune system itself. Ironically, when the immune system reacts to a microorganism, it is, in effect, encouraging the microorganism to mutate.[18] In this regard, those antigens that elicit the strongest immune response will be subject to the greatest immune selection pressures.

2.3.2. Regulation of complement activity

One of the most important components of immunity is the *complement system*. This is a series of some thirty or so substances in blood serum that become activated in a series of reactions during an immune response (known as a 'complement cascade'). This process can be activated by microbial substances during innate immune responses, but also by antibodies in adaptive responses.

This is a further example of the importance of system balance. Insufficiencies in key components of complement would result in a devastating outcome with regard to certain infectious diseases, despite the use of antibiotics or other chemotherapeutic agents. On the other hand, unrestrained complement activation would cause severe damage to bystander cells. In a healthy body, complement activity is held in check by a variety of regulatory factors, known as regulators of complement activation (RCA).[19]

Members of the poxvirus, herpesvirus and retrovirus families produce homologues that mimic RCA proteins and are thus able to escape complement action.[20] The smallpox virus *Variola major* causes a serious, virulent infection in humans, while the virus that is used for vaccination against smallpox, vaccinia virus, usually causes only a very mild or even unapparent infection, at least in individuals with an intact immune system.

A component of the smallpox virus that may contribute to its pathogenicity or ability to cause disease is the smallpox inhibitor of complement enzymes (SPICE). SPICE has the ability to inactivate one of the key complement components (human C3b) that serves to induce the innate immune process by which cells engulf material which is eventually digested, destroyed or killed. By inactivating the complement activity, a vital area of innate immunity would be disabled. Vaccinia virus also has a complement regulatory protein (called vaccinia virus complement control protein, VCP), which is, however, much less effective (100-fold less) than SPICE. In a recent report, researchers mutated the VCP gene of vaccinia virus to have the same nucleotide sequence as SPICE.[21] The recombinant mutant VCP proved to be much more efficient than normal VCP in inactivating complement in a test tube reaction. Although the researchers did not actually outfit vaccinia virus with this mutated gene, the work was only one step away from this manipulation. Presumably, vaccinia virus with the mutated gene would be much more pathogenic.

2.3.3. Regulation of cytokine activity

As previously mentioned, interferons are cytokines produced by cells to protect them from viral infection, and anti-interferon strategies are a part of the immune evasion repertoire of most viruses. These mechanisms include the production of soluble versions of interferon receptors, which act as decoys. These decoys bind and inactivate interferons before they reach their 'destination' – normal, membrane-bound receptors.[22]

The activities of the proinflammatory cytokines IL-1β, TNFα and IL-6 have been referred to above. Other cytokines, such as IL-2, IL-4, IL-10 and IL-12 are essential in directing the activities of different arms of the immune system. One of the most interesting evasion mechanisms identified in recent years is the mimicry of cytokines and cytokine receptors by large DNA viruses (herpesviruses and poxviruses). Cytokine homologues might redirect the immune response for the benefit of the virus, for example by suppressing the anti-viral activity of cytotoxic T cells. Alternatively, viruses that infect immune cells might use these homologues to induce signalling pathways in the infected cell that promote virus replication.[23] Furthermore, soluble cytokine receptors made by the virus might neutralize cytokine activity before the cytokines could react with their normal, membrane-bound receptors.

2.3.4. Inhibiting programmed cell death

A further immune evasion strategy includes the production of a variety of viral inhibitors of cell death (apoptosis), the so-called programmed cell death. In this regard, apoptosis can be viewed as a response to limit the intracellular propagation of viruses. The immune system recognizes a cell that has been infected by a virus through the presentation by that cell of fragments of viral proteins bound to MHC molecules on the surface of the cell. As stated above, unlike a B lymphocyte, a T lymphocyte will only recognize a virus that is attached to a MHC molecule. This recognition leads to the activation of cytotoxic T lymphocytes which attack and kill the cell through the induction of apoptosis.

Some viruses can cause the suppression of the production of MHC molecules. This would mean that viral antigens would not be bound to MHC molecules and could not be recognized by T cells. The cell and therefore the virus production factory would be protected from cytotoxic T lymphocyte destruction.[24] Alternatively, viruses such as cytomegalovirus induce the expression of a certain type of MHC molecule that can bind a receptor on the surface of natural killer cells, inducing suppression of the activity of these cells that are normally an important component of innate immunity.[25]

3. Dual-use aspects of biomedical research

There are four categories of manipulations or modifications of microorganisms and their products that have been the subject of discussion since the onset of the development of genetic engineering: (1) the transfer of antibiotic resistance to microorganisms, (2) modification of the antigenic properties of microorganisms, (3) modification of the stability of the microorganism toward the environment and (4) the transfer of pathogenic properties to microorganisms.[26] All of these four types of manipulations are being carried out daily in research programmes that have legitimate and basically peaceful aims, such as the elucidation of the mechanisms of microbial pathogenesis. This research is essential for developing better means of combating infectious diseases. At the same time, these techniques can be misused to produce new types of biological agents that could be used as weapons. In order to focus more directly on the dangers involved, specific examples of work from the recent literature that has produced dangerous microorganisms which are able to evade vital immune mechanisms will be examined.

3.1. Tranfer of virulence genes to *Bacillus anthracis*

This work was carried out by a group of investigators and involved the transfer of haemolytic toxin genes from the bacterium *Bacillus cereus* to not only vaccine, but also virulent strains of *Bacillus anthracis*.[27] The toxin in question is called cereolysin, whose haemolytic properties are apparently determined by phospholipase enzymes that damage cell membranes, thus killing the attacked cells. This action can be visualized by using red blood cells as an indicator, because the membranes of these cells are very susceptible to disintegration after damage has been caused and they will readily lyse (therefore the description 'haemolytic'). According to the authors, the study was undertaken to investigate changes in immunogenic properties of vaccine strains in connection with haemolytic characteristics. An unexpected result was obtained with one virulent strain of *Bacillus anthracis* that received the cereolysin genes from *Bacillus cereus*: the normal *Bacillus anthracis* vaccine strain failed to protect against a challenge infection of hamsters with this engineered virulent strain, while a vaccine strain which was also transfected with the cereolysin genes did afford protection. Thus, tranfer of additional virulence genes to a pathogenic *Bacillus anthracis* allowed it somehow to evade the defences provided by the normal vaccine. The mechanism of the evasion is not clear.

It cannot be denied that knowledge gained by these experiments concerning the behaviour of the engineered strains might be useful for BW defence or even medical purposes. However, this work has to be classified as being of a highly dual-use character in the light of the outcome of the experiments, that is, the development of a strain of *Bacillus anthracis*, a prototype potential biological weapon that is able to evade protection provided by known vaccines.

3.2. Accidental creation of a 'killer' mousepox virus

The potential dangers that may be associated with biological research are particularly evident in recent studies in the area of immunology. The headlines in the journal *New Scientist* proclaimed 'Disaster in the making. An engineered mouse virus leaves us one step away from the ultimate bioweapon.'[28] The report was about experiments carried out by Australian researchers who tried to make mice infertile, as a model for controlling rodent populations.[29] The experimental strategy was to incorporate a gene for the production of a protein that is found on the surface of egg cells of the mouse into the genome of a mousepox virus, against which the mice used in the experiment were resistant. When the mice were infected with the recombinant virus, the egg cell protein was over-produced, and an antibody response to that protein was mounted, which was supposed to cause infertility in the mice. Indeed, the expected antibody response occurred, but it was short-lived.[30] In order to boost these antibody responses and prolong their effects, another gene was introduced into the mousepox virus genome.[31] This gene was to direct the production of a substance called interleukin 4 (IL-4), which is known to enhance antibody-type immune responses. However, IL-4 also suppresses the activation and expansion of another type of T-lymphocyte (Th1) that provides essential help to cytotoxic T-lymphocytes (CTLs) needed to fight viral infections. When mice were infected with the recombinant virus, the IL-4 produced did boost antibody responses to the mouse egg protein, but at the same time it also suppressed the activation of CTLs. As a result, most of the mice (60 per cent) died, even though they were supposed to be resistant to the virus.[32]

The mousepox virus is not infective for humans. However, there is some concern that the same manipulations might be performed on a pox virus that does infect humans, with devastating results.

This work has been continued by Mark Buller, a professor at Saint Louis University. Buller constructed a recombinant mousepox virus containing the IL-4 gene that was even more deadly than the one made by Jackson and co-workers. By placing the IL-4 gene in a region of the virus genome that was dispensable and by optimizing its expression, 100 per cent mortality of the resistant mice was achieved with the recombinant virus construct. Buller's stated motivation was to explore possible prophylactic and therapeutic defences against such an agent. However, vaccination or treatment of mice with the antiviral substance cidofovir plus antibodies to IL-4 still did not protect them adequately against a challenge with the highly virulent mousepox constructs.[33]

Now he has apparently gone one step further with plans to alter the cowpox virus, which can infect humans, in a similar way.[34] Buller asserted, however, that this virus would only be lethal in mice and not in humans, because he used the mouse IL-4 gene, which is specific for only the mouse immune system. The head of the Australian research team, Ian Ramshaw, maintains that there is no reason to do the cowpox experiments. He further cautions that while viruses containing the mouse IL-4 gene should not be lethal in humans, recombinant viruses can have unexpected effects.[35] Indeed, it has been pointed out[36] that these experiments fall into several categories listed by the National Research Council Report as being research of particular concern.[37]

3.3. Potentiation of the virulence of influenza A viruses with genes from the 1918 pandemic virus ('Spanish' influenza)

The 1918 influenza pandemic was the most catastrophic flu outbreak, causing more than 20 million deaths worldwide. Recently, a team of researchers headed by Jeffrey Taubenberger at the Division of

Molecular Pathology, Armed Forces Institute of Pathology in Rockville, Maryland, reported that they had potentiated the virulence of a normal strain of influenza virus by incorporating into its genome genes from the 1918 pandemic virus. The two genes that were most decisive in determining virulence were the genes encoding the haemagglutinin (HA) and the neuraminidase (NA) of the 1918 strain.[38] Several years earlier these researchers had begun to isolate genes from this virulent strain and sequence them using newly developed molecular biological approaches.[39] The reasons given for undertaking this research included the possibility of carrying out a molecular analysis of the pathogenic mechanisms of the 1918 virus as well as testing the effectiveness of current antiviral drugs against a strain of this virulence. They report that recombinant viruses harbouring the 1918 HA and 1918 NA were inhibited effectively in both tissue cultures and in mice by the NA inhibitors zanamivir and oseltamivir.[40]

These experiments illustrate the absolute dual-use dilemma of research in the biotech sector. While such experiments create microorganisms that pose a greater risk than do normal ones and the advisability of their undertaking is certainly open to question, there may at the same time be benefits from such research. In this respect, the National Research Council report states that 'even experiments that have the greatest potential for diversion to offensive applications or terrorist purposes may also have potentially beneficial uses for public health promotion and defense'.[41] This points up the difficulty in imposing blanket prohibitions on certain research activities from the start, but clearly emphasizes the need for oversight of research of BW relevance. These aspects will be discussed further in the last chapter. With the last two examples cited, it can be seen how natural viruses of limited pathogenicity can be restructured to mimic the effects of agents of much higher virulence.

3.4. Creating viruses using molecular biology and information technology

The use of modern techniques of molecular biology and information technology to create viruses that have been the subject of recent reports is of particular concern, such as the creation of the polio virus from scratch[42] or the generation of a bacterial virus within two weeks using synthetic oligonucleotides.[43] These reports have in any case initiated a debate about whether unclassified research that might conceivably be misused by terrorists should be openly published.[44] These viruses are albeit simple in comparison to most viruses of BW relevance, such as the poxviruses.

There is a great deal of interest in finding ways to manipulate poxvirus genomes in vitro. Poxviruses have genomes that are composed of linear double-stranded DNA molecules that are resolved from transient head-to-head or tail-to-tail structures called concatemers during replication. The introduction of new genes into the vaccinia virus genome is usually carried out by homologous recombination in mammalian cells, which is rather inefficient. Also, time-consuming selection procedures are required. A way to facilitate manipulation of this complex genome has been reported.[45] A method of cloning the entire genome of vaccinia virus (VAC) as a continuous molecule in a plasmid (circular DNA molecule that can be replicated in bacteria) taking on the form of a bacterial artificial chromosome (BAC) was developed. This VAC-BAC could thus be stably propagated in the bacterium *Escherichia coli* and subsequently converted into infectious virus in mammalian cells. This system can greatly facilitate genetic studies in poxviruses, but at the same time it is decidedly dual-use.

4. Future threats

4.1. Vulnerability of the immune system to modulation with bioregulators

In addition to immune evasion by pathogens, there has to be a great deal of concern about the possibility of modulating immune responses in a negative way with bioregulators that are not microorganisms, but rather substances found normally in the body that regulate biological processes.

The inappropriate production of proinflammatory cytokines can be taken as an example of malign use of bioregulators. Enhancing the proinflammatory cytokine production by using PAMPs to engage Toll-like receptors on the surface of macrophages could at the very least lead to sickness behaviour, which is characterized by fever, drowsiness, lethargy and loss of appetite (normally signs that the immune system is 'kicking in').[46] However, if the proinflammatory cytokines are produced in particularly large amounts, this could lead to autoimmunity, or eventually even to shock and death.[47] On the other hand, inhibiting the production of these cytokines by using bioregulators that can negatively regulate their synthesis might result in a lack of innate immune protection.

A second example of modulation of immune responses with bioregulators concerns 'super-antigens'. The immune system is particularly vulnerable to attack by certain super-antigens. Normally, less than 0.01 per cent of B or T lymphocytes respond to a particular antigen. In contrast, a number of super-antigens has been described that can react with a significant proportion of T lymphocytes (between 5–25 per cent).[48]

For example, the bacterial product *Staphylococcus* enterotoxin B (SEB) is a biological agent that also falls into the category of a potential chemical weapon. This toxin was on the US list of favoured anti-personnel agents as early as 1949[49] and was apparently weaponized by the US Army prior to the negotiation of the Biological and Toxins Weapons Convention (BTWC).[50] It has also been the subject of extensive research in the biomedical literature. SEB acts as a super-antigen in that it can activate a large proportion of T lymphocytes to produce excessive amounts of cytokines, which can cause systemic reactions including inflammation, fever, widespread blood clotting and shock.[51]

Recently, a B cell superantigen has been described that can bind up to 50 per cent of the B cell population,[52] resulting in an increased rate of apoptosis or death of the bound cells. The researchers of this B cell super-antigen are considering putting it to therapeutic use.[53] They are engineering the antigens to achieve higher binding affinities and different specificities in order to specifically target malignant B cell populations such as lymphoma and leukaemia.

4.2. Targeted delivery systems: gene vectors and immunotoxins

Targeted delivery systems have to be characterized as being strongly dual-purpose. While they may be potentially very useful in vaccine and gene therapy, they can also serve as delivery vehicles for toxins or bioregulators in a negative way.

These delivery systems are components that allow an activity to be targeted to a particular site in the body where its activity is desired. An example of such a sytem are viruses that are used as vectors to transfect a foreign gene into cells for the purpose of immunization or for gene therapy. Vaccinia virus has been investigated for these purposes because of its large genome, which can carry several foreign genes at once, and its effectiveness as a vaccine.[54] There has been a great deal of work in recent years on the possibility of using adenoviruses as gene vectors. Adenoviruses, the causative agents of acute respiratory and ocular infections, are widespread in nature. They are among the most efficient gene delivery vehicles in use today. They have a broad host range and do not integrate into chromosomal DNA, which reduces the risk of mutations via insertion. Retroviruses, by contrast, integrate randomly into the genome of the host. Furthermore, adenoviruses can be produced at high titers (up to 10^{10} per milliliter)[55] and they also have a carrying capacity of up to 40 kb of insert DNA. Alternatively, the development of adeno-associated viruses as vectors for gene delivery seems promising, as these viruses are defective by nature and have thus never been shown to have any pathogenic effects in humans.[56] However, latest investigations have shown that these viruses do indeed integrate into the host genome more frequently than presumed, so that there are still serious safety concerns with these vectors.[57]

Another prime example of a targeted delivery system are immunotoxins. These are molecules that contain the antigen binding specificity portion of an antibody molecule coupled to a toxin molecule. The aim is to target the toxin activity to specified cells, such as tumour cells; in this case, the antibody specificity is directed against tumour cell antigens.[58] The antibodies are usually monoclonal antibodies produced in mouse cells. The toxin portion is usually the toxic fragment of a toxin molecule minus its binding component. Without its binding component, the toxic fragment cannot dock onto target cells, penetrate into the cell, and exert its toxic effects. When coupled to an antibody that can bind specified target cells, the toxic fragment is directed to those cells and acts on them. The inactive toxic fragment is generally produced by genetic engineering in bacteria or animal cells and it is coupled to the monoclonal antibodies by a chemical reaction in vitro. If the immunotoxins are to be used for human therapy, the mouse antibodies can be 'humanized' by fusing the antigen binding portion of the mouse antibody molecule to the remaining portion of a human antibody molecule via genetic engineering.[59] This is done to minimize the induction of an immune response against the foreign protein immunotoxin before it has time to work.

The toxins that have been used to produce immunotoxins include ricin, *Shigella* toxin, and diphtheria toxin. A number of clinical trials using immunotoxins have been completed, while others are still going on. To date, results have been promising in leukaemia and lymphoma patients, but responses in patients with large tumours have been disappointing. Clinical trials have highlighted several problems with the administration of immunotoxins, including vascular leak syndrome, liver toxicity, and toxicity to tissues of the central nervous system; current research is directed towards increasing the tolerability and targeting specificity of immunotoxins.[60] In this regard, hybrid genes that can be used to produce antibody-toxin combined molecules (called fusion proteins) in bacteria have been designed, using only the antigen-binding portion of a single chain antibody molecule attached to the toxic portion of a toxin molecule. Alternatively, a fusion protein could be produced that consists of an antigen (ligand) for a receptor fused to the toxic fragment of a toxin molecule. New strategies to reduce immune reactions against the immunotoxins have been developed.[61]

Aerosolization of vectors (dispersing them as fine droplets or particles suspended in the air) carrying foreign genes could represent an effective delivery system, especially if the vector is a virulent microorganism, as most infections begin at the mucosa. If the vector is not a microorganism, such as in the case of fusion proteins or immunotoxins, successful delivery by the aerosol route would depend greatly upon the physical-chemical properties of that vector. In its medical research work on endogenous bioregulators the US Army has for example experimented with the aerosolization of the hormone insulin and the cytokine interleukin-1. It was reported that these biochemical agents were effective in aerosol form in basic pulmonary absorption studies.[62]

There are still many technical problems involved with the use of targeted delivery systems that would serve to limit their application. However, there is a tremendous interest in developing these systems further for therapy purposes, so that we can expect great advances in this area in the near future that would be of particular concern regarding misuse.

4.3. Immunization with plant foods

There is at present a great deal of interest in developing vaccines as plant foods. This involves the transfer of a gene encoding the antigen of interest into the genome of plants, with subsequent expression of that gene and biosynthesis of the antigen in the plant tissues. Eating the plant tissues would then deliver the antigen to the gut, where it would be taken up by special epithelial cells of the small intestine (M cells) and transferred to the underlying lymphoid tissues, resulting in an immune response to that antigen. There would be several advantages of inducing an immune response in this way, including increased safety, economy and stability of the vaccine, as well as the prospect of inducing mucosal immunity (to localize immunity at mucous membrane sites, where most infections begin).[63]

Although the advantages of edible vaccines are many, there are numerous technical and immunological hurdles that have to be overcome in order for them to be practical. One of the first is the avoidance of degradation of the antigen in the digestive tract. Even if the antigen would survive this degradation, oral tolerance mechanisms would have to be overcome. This is a type of tolerance to antigens administered orally, which prevents immune responses to the microorganisms residing in the intestine or to protein antigens aquired continually in food. This tolerance might be overcome if the vaccine is administered with a mucosal adjuvant (a special type of immune response booster) or if the antigen is in the form of particles. Furthermore, oral immunization usually requires multiple doses in larger amounts than antigen administered over parenteral routes; responses are weak, unreliable and also shorter lived.[64] Indeed, results to date show that immunization with plant foods is in some cases possible, but the responses are usually modest and appear only after more than one dose.[65]

One of the most successful preparations to date is that of an edible vaccine for hepatitis B.[66] Volunteers who had been previously immunized parenterally (by injection, not by mouth) with the licensed, recombinant hepatitis B vaccine in yeast were given three doses over a period of 28 days of the hepatitis B antigen (HBsAg) as a recombinant protein in potatoes. The doses consisted of 100 to 110 grams of the potato that were ingested by the volunteers. Nine of the 17 volunteers responded with significant antibody production over those values measured before they ingested the potatoes. The serum antibody titers increased up to 56 fold (range 1.3–56 fold) in these individuals. This showed that the plant vaccine without any adjuvant could produce a significant response in individuals that had been immunized previously with the commercial, licensed HBsAg vaccine. These studies did not, however, test the response of subjects that had not been previously immunized. The success of this particular vaccine is no doubt due to the fact that the recombinant protein is one that can assemble into aggregates. This property of HBsAg is well-known and has been responsible in the past for its success as a recombinant protein parenteral vaccine in vehicles such as yeast cells.

This discussion serves to illustrate that immunization with plant foods is by no means readily achievable. In this regard, it is unlikely that these techniques can be used successfully in the near future in malign ways, e.g. for vaccination of unaware populations, thus forcing upon them an involuntary immunity or marking them as possible targets. Nevertheless, there is great interest in developing such vaccines for peaceful use and improvements are actively being sought,[67] which will no doubt be achieved. Work is focusing at the moment on producing plant vaccines consisting of an antigen fused to a protein that can target the antigen to the mucosa (gut), in order to produce a local immunity on the mucous membranes, where most infections begin.[68] Therefore, developments in this area should be closely monitored.

4.4. Vulnerability of the immune system to modulation after immunization

Activation of the immune system in response to an infection is a vital step in countering the threat posed by the causative agent. Nevertheless, activation of components of the immune system is invariably associated with the enhanced production or exposition of predictable markers, that could serve as targets for the delivery of a biological weapon to those sites.

B and T lymphocytes are produced during development and prior to encountering antigens by rearrangement of genes to yield an enormous number of clones, each expressing a unique receptor recognizing a particular antigen epitope and thus being able to respond to that antigen.[69] Initially, only a small subset of these clones is able to recognize any one particular antigen. For example, it can be assumed that normally less than 0.01 per cent of the lymphocyte repertoire can recognize a particular protein antigen.[70] To generate effective immunity, these naive or resting B cells and T cells must undergo clonal expansion in response to an antigen challenge in order to amass the numbers required to counter an infection. Depending on the strength of the challenge and the type of antigen, the naive lymphocytes are activated and then driven to proliferate in approximately 10–20 turns of the cell cycle, before they cease proliferation and proceed into a phase of differentiation, after which they are able to execute their functions. This represents a considerable expansion of antigen-specific lymphocytes in response to immunization, especially when a vaccine is given in several doses over period of time.

These expanded clones of B and T lymphocytes carry receptors specific for a particular antigen (B cells) or a fragment of that antigen bound to MHC molecules (T cells). These cells have an enhanced vulnerability, for example, to being targeted with constructed toxins as discussed earlier (targeted delivery systems). For delivery to B cells, a delivery system might be a fusion protein consisting of the specific antigen (against which the B cells are directed) fused to the toxic chain of a toxin molecule (such as the A chain of ricin or diphtheria toxin). However, since B cells release antibodies to the antigen, the construct might be neutralized and cleared by these antibodies before it could do much damage. T cells might be a more vulnerable target, as they do not secrete their antigen receptors, but the delivery system containing the toxin would have to be constructed in a way as to include the specific antigen fragment bound to MHC molecule epitopes in order for it to be recognized and engaged by the T cell. This would be a tall order at present, but new studies are providing greater insight into the fine points of the recognition of antigen presented by MHC molecules to T cells[71] that could make this approach more cause for concern in the future.

In addition to the expansion of specific antigen receptors, immunization also up-regulates the exposition of an array of molecules on the surface of lymphocytes and macrophages. Prominent ones include MHC molecules on lymphocytes and macrophages, CD40 on B cells and macrophages, or CD28 and CD40L on T cells. All of these would be vulnerable to attack for example with immunotoxins consisting of antibodies to these surface components bound to the toxic chain of a toxin.

5. Summary

In this chapter, the dual-use dilemma of modern biotechnology has been viewed within a broader scope of consequences by focusing on biological systems as the target of potential malign intent, using the immune system as an example. The possibility of the perturbation of this system not only with microorganisms designed to evade immune defences, but also with bioregulators that can profoundly affect its function, raises the dual-use dilemma to a higher order of concern. In this regard, innate immune mechanisms, which represent the essential first line of defence against infections, are particularly vulnerable. If key functions of this system are malignly attacked, the battle against infections is lost from the start. This system is additionally vulnerable to modulation of proinflammatory cytokine production. An overproduction of these cytokines could lead to debilitating sickness behaviour and even to shock or death. Furthermore, the mousepox experiment clearly shows what devastating effects modulation of adaptive immunity by cytokines can have on immune defence mechanisms. The possibilities for misuse of knowledge about immune mechanisms and their regulation for disrupting the function of the immune system are intricately involved with dual use aspects of targeted delivery systems. Although there are technical problems involved with the deployment of substances to the desired targets by delivery systems, there is great interest in developing these for use in cancer, gene and drug therapy, so that improvements in this area will be seen in the near future.

5

Behaviour Under Control: the Malign Misuse of Neuroscience

1. Introduction

Only in the last few centuries has the link between the brain and behaviour become clear, and only at the end of the nineteenth century was it demonstrated that the nervous system was made up of billions of separate nerve cells or neurons. We now know that during evolution complex networks of such neurons have developed in order to effect certain behaviours. Whilst the neurons of the central, peripheral and autonomic nervous systems vary enormously in form and function, they can be classed into three broad groups: sensory neurons which convey information into the central nervous system; effector neurons which carry information out of the central nervous system to muscles and other effector organs; and interneurons within the central nervous system which link the sensory and effector neurons and also have links with one another.

Information is conveyed *within* individual neurons by electrical means – generating nerve impulses which can be recorded and displayed on an oscilloscope. In the twentieth century it was shown that information is conveyed *between* neurons by chemical means. When a nerve impulse (an action potential) travelling along the long extension or axon of a neuron arrives at a junction (or synapse) with another neuron, it causes the release of a neurotransmitter chemical from the presynaptic cell. This chemical affects the electrical properties of the postsynaptic neuron through its interaction with specialized receptor proteins embedded in the surface membrane of the postsynaptic cell. It has been shown that there are numerous kinds of neurotransmitter chemical which, depending on the specific receptors involved, can either cause an electrical change which enhances the possibility of an action potential occurring in the postsynaptic cell, or, alternatively, decreases that possibility. Various chemical mechanisms ensure that the neurotransmitter is cleared from the synaptic area so that its effect does not persist and so that another action potential in the presynaptic neuron can exert its effect in turn.

This then is the basis for modern insights into how the brain – and therefore behaviour – can be manipulated by chemical means. Clearly, as our understanding of the neuronal circuits underlying specific behaviour increases, and we understand more about the neurotransmitters functioning in such circuits, we have more chance of helping people who are suffering from various malfunctions of the nervous system (mental illnesses). It has to be accepted, however, that such information may be misused by those with malign intent.

Thus in the early years of the east–west Cold War, following the serendipitous discovery of chemical agents (drugs) that could help people with severe mental illnesses, the military took an interest in many different means of chemical incapacitation. The original 1970s study, *CB Weapons Today*,[1] from the Stockholm International Peace Research Institute (SIPRI), states that the United States Army Chemical Corps drew attention to at least a dozen mechanisms in the late 1950s and it gives details on, for example, hypotension, emesis and disturbance of body temperature and further, lists loss of balance, muscular hypotonia, muscle tremors, and 'many different psychotropic effects' on the central nervous system produced by tranquillizers, sedatives, anti-depressants and psychotomimetics. At that time the SIPRI authors argued that there was too little knowledge of the workings of the central nervous system for such central effects to be used successfully to incapacitate. It has been argued since that the genomics revolution of the 1990s will have made a significant difference because the structure of the receptor sub-types affected by relevant neurotransmitter chemicals has increasingly become known.[2] It would thus be much easier to design chemical incapacitating agents today to achieve specific effects.[3]

The genomics revolution, however, clearly has much more profound implications for our understanding of biological systems – including those of the central nervous system. As a recent major review noted, the possibility of a systems-level understanding is gaining importance because:

> progress in molecular biology, particularly in genome sequencing and high-throughput measurement, enables us to collect comprehensive data sets on system performance and gain information on the underlying molecules.[4]

From this viewpoint, what has changed is not just that the genomics revolution has enabled the elucidation of the receptor sub-types involved in central nervous system circuits, but that the molecular mechanisms in whole control systems governing particular behaviours may be elucidated. Another recent review suggested that our increasing ability to understand complex signalling relationships in the central nervous system will enhance the possibility of finding new therapeutic drugs. The review argued that the need for such an approach was necessitated by the current difficulties in dealing with a major illness like depression.[5] Yet despite this shift, the problem of unpleasant side-effects and thus patient non-compliance with treatment regimes remains. The attraction of a systems approach combining the expertise of engineers, biologists and mathematicians,[6] while in addition building on the growing capabilities of neuroimaging, is undoubtedly powerful. How far then do we understand how behaviour is controlled by systems within the brain? A good deal more progress in micromeasurement technology will clearly be required before the kind of hypothesis-driven modelling, prediction and testing to which systems biologists aspire is completely possible,[7] but great progress is evident in some areas, and brain circuits and neurotransmitter/neuroreceptor functions are being elucidated as the brief following example illustrates.

1.1. Noradrenaline/arousal

There has been a clear military interest in manipulation of the noradrenaline neurotransmitter system in relation to the arousal level of the central nervous system for some time.[8] That interest apparently continues; one recent report suggesting that drugs affecting the system were 'appropriate for immediate consideration as a non-lethal technique'.[9]

Noradrenaline is a small-molecule, classical, neurotransmitter which has an unusual distribution in the mammalian central nervous system.[10] The major noradrenaline cell group is the locus coeruleus (LC group). The locus coeruleus contains a surprisingly small number of neurons, some 20,000 in the rat. However, these neurons have axons which branch profusely in the brain, so noradrenaline acts as a neurotransmitter in many different brain regions.

Not surprisingly, noradrenaline transmission is involved in many brain functions. As noted in a recent major review:[11]

> the NE (noradrenaline) system plays an integral role in the modulation of neuronal function and behaviour, including arousal, vigilance, learning and memory.

Extensive studies in rats, cats and monkeys have elucidated much of how this system functions in arousal and vigilance.[12] Brain areas known to be involved in attention receive particularly dense innervation from the LC noradrenaline neurons. Moreover, spontaneous (tonic) activity of LC neurons in the rat varied consistently with the animal's behavioural state – firing most rapidly during waking, less during slow-wave sleep and rarely in paradoxical sleep (equivalent to rapid eye movement (REM) sleep in humans). Similar results were obtained with cats. In monkeys and rats a lower rate of firing was found in the waking animal when it was engaged in automatic/vegetative activities such as grooming and drinking. Conspicuous environmental stimuli in many modalities evoked an enhanced (phasic) response in LC neurons and evoked a behavioural response orientated to the stimulus.

Though the detailed operations of this noradrenaline system are complex and though it may have greater involvement in vigilance than just with arousal levels, one simple point stands out and is of particular interest to those with malign intent. The noradrenaline neurons have autoreceptors, both in the cell body region and in their terminal regions, which respond to noradrenaline and bring about a reduction in the neuronal output by reducing their firing rate and thus in neuronal transmitter production.[13] In short, they are inhibitory autoreceptors – production of noradrenaline limits its own production. Thus it has been found that drugs which affect such receptors in the same way as the noradrenaline natural transmitter (agonists) will have the same effect, and if given in sufficient quantity will stop the LC neurons' activity and put the animal (or human) to sleep. Not surprisingly, one military report in the early 1990s noted that such compounds:

> have been considered to be ideal next generation anesthetic agents which can be developed and used in the Less-Than-Lethal [Non-Lethal] Technology Program.[14]

And researchers had already discovered chemical agents like dexmedetomidine which had such effects and were attempting to modify their chemical structure in order to increase their specificity.[15] It is obviously in that regard that increasing understanding of receptors and their sub-types becomes important.

Receptors which respond to noradrenaline are called adrenoceptors (as they also respond to adrenaline). All of these receptors are slower-acting G protein-coupled receptors (GCPRs) which are located on cell membranes. It has been known for over 50 years that there are two broad classes of such receptors and these are termed α- and β-adrenoceptors, and it has been clear since the 1970s that some adrenoceptors are located presynaptically as well as postsynaptically. The coming of molecular biology allowed the genes for six human α-adrenoceptors to be identified (α_{1A}, α_{1B}, α_{1D} and α_{2A}, α_{2B} and α_{2C}).[16]

The α_2-adrenoceptors function as the presynaptic autoreceptors and, with the advent of the capability to produce mice with different genes knocked out, it has been possible to show that the α_{2A} and α_{2C} sub-types operate as the presynaptic inhibitory receptors in mice. Furthermore, it is clear that approximately 90 per cent of these receptors are of the α_{2A} subtype.[17] Studies are now in progress to elucidate the nature of numerous polymorphisms in the nine different subtypes of human α- and β-adrenoceptors and the role, if any, these play in various diseases.[18] Dexmedetomidine was originally released as a veterinary sedative-analgesic and in the United States in March 2000 as an anaesthetic for sedation of intensive care patients. So the knowledge of this system and how to modify it has beneficial outcomes.[19] It is certainly also possible that such knowledge could be misused, particularly as understanding and capabilities grow.

It might be argued, of course, that whilst our understanding of the nervous system has increased since the 1970s, the ability to sedate people by targeting the LC noradrenaline system does not in fact indicate a step change in real capabilities. After all, as the original 1973 SIPRI study noted:

> Several bacterial endotoxins are amazingly potent fever-inducers in man, effective at submicrogram dosages, in addition to producing other incapacitating effects, such as pain, vomiting and sensitization to other materials.[20]

So it might be argued that those wishing to use neuroscience for malign purposes would need much more impressive evidence of current penetration into the workings of the central nervous system before expending much effort in such a direction.

2. Investigation of the possibilities

There are many examples in the current literature that amply demonstrate how different are the present capabilities in comparison to those of the 1950s. The following are therefore only illustrative of the dangers.

2.1. Post-traumatic stress disorder (PTSD)

According to the standard *Diagnostic and Statistical Manual of Mental Disorders* (4th Edition of 1994, DSM-IV) you have post-traumatic stress disorder if:

> You have been exposed to a horribly traumatic event that made you feel extremely fearful, helpless, or terrified.
>
> You keep reexperiencing the event in different ways, such as upsetting memories or nightmares; flashbacks that it is happening again; or having a severe reaction whenever you are exposed to anything that reminds you of it.
>
> You avoid things that are associated with the traumatic event; cannot remember the details of what happened; feel detached from everyday life; or feel like you will never have a normal life again.
>
> You are jumpy and hypervigilant, having trouble sleeping, have angry outbursts, or have trouble concentrating.
>
> These symptoms persist for at least a month and cause either severe distress or problems with school, work, or other people.[21]

During evolution the human species has evolved mechanisms to ensure that dangerous events are well remembered for the obvious good reason of either avoiding such events, or taking great care about them, in the future. If this response gets out of hand we call it post-traumatic stress disorder (PTSD), and it clearly causes great distress to those who suffer from it. There is every reason for trying to understand how it comes about and to find better ways of dealing with it.

It is not too difficult to discern that PTSD involves at least two components: learning and memory. These concepts may be defined in this way:

> The acquisition of reproducible alterations in behaviour as a result of particular experiences is *learning* ... *Memory* is the storage of the altered behaviour over time.[22]

We are clearly dealing here with learning about aversive events and consolidation of the memory of such aversive events. The basic elements of the system for dealing with fearful events is built into all mammals. Thus, if we hear a large explosion we will, like the rat, exhibit a startle response and freeze momentarily before the flight-or-fight response kicks in.[23] Therefore, it is possible to gain much insight into the human fear system from investigations of those in other mammals like the rat.

It is relatively easy to study the impact of fear on the rat through what is called classical fear conditioning (after Pavlov). The rat is repeatedly subjected to a sound (which it does not fear) followed by a mild electric shock (to which it does react with fear). Soon it learns to react to the sound alone in anticipation of the shock. Investigators like LeDoux knew that sound picked up in the ear is processed in the auditory mid-brain, then the auditory thalamus and finally in the auditory cortex (the highest level). Surprisingly, when lesions were made in the auditory cortex it was found that rats could still associate the shock and sound and were still reacting with fear to the sound alone. The auditory cortex is clearly not required to support such behaviour.

Further investigation showed that lesions in either of the sub-cortical levels (auditory thalamus and auditory mid-brain) eliminated the fear conditioning. The information was obviously being processed somewhere beyond the thalamus, but not in the auditory cortex, in order that the fear reaction occurred. This location was found to be the amygdala – which was not too surprising since the amygdala has been known for years to be important in emotional responses. Ledoux explained as follows:

> The low road, of the thalamo-amygdala pathway, is a quick and dirty system. Because it doesn't involve the cortex at all, it allows us to act first and think later ... We freeze first, and that gives us a few seconds to decide what to do: Run away? Hold still? Try to fight?

If we are in a wood and see a stick that might possibly be a snake we are better reacting immediately as if it were indeed a snake. However:

> The cortex – the high road, so to speak – also processes the stimulus, but it takes a little longer.

While the amygdala pathway prepares for action the cortex pathway is simultaneously processing the information, and if it decides that what is seen is actually a stick and not a snake little effort is wasted as it can switch off the emergency response. That is fine as far as it goes, as one study noted:

> There is now extensive evidence suggesting that the amygdala is involved in the effects of attention and reward ... and that the amygdala may be a locus of the neural changes underlying the acquired association of cues with emotional responses, especially the autonomic and motoric responses elicited by fearful stimuli.[24]

However, the passage continued:

> In addition, there is a strong consensus that the amygdala is involved in mediating the effects of emotional arousal on memory. Findings of many studies indicate that the amygdala mediates the consolidation of long-term explicit memories of emotionally arousing experiences by influencing other brain regions involved in memory consolidation.

It is this second process of memory consolidation that is surely of more interest in relation to PTSD.

A variety of evidence shows that the amygdala is not the site of long-term memory.[25] So something more complex is happening than the amygdala operating in isolation. In fact, the system is very complex and is, as yet, far from completely understood. Enough is known, however, to suggest that systems biologists will decipher it rather quickly.

It is well known that under stress the hypothalamus secretes corticotrophin-releasing hormone (CRF) into the pituitary portal blood supply and this causes the anterior pituitary gland to produce adrenocorticotrophic hormone (ACTH) which, in turn, enters the general circulation and causes the adrenal cortex to secrete glucocorticoids. At the same time the sympathetic nervous system activates the secretion of adrenaline from the adrenal medulla. These two agents, glucocorticoids and adrenaline, have significant effects on the mobilization of the body, but they also affect the functions of the amygdala. First, there is considerable evidence that adrenaline, which does not pass the blood-brain barrier, still has an indirect impact. β-adrenoceptors on afferents in the vagal nerve send information into the central nervous system which leads to enhanced noradrenaline neurotransmitter output in the amygdala. Perfusion of the amygdala with β-adrenoceptor agonists after training enhances memory consolidation and antagonism of β-adrenoceptors blocks the enhancement. It has been concluded that such findings:

> provide strong support for the hypothesis that noradrenaline release an the amygdala plays an important, possibly critical, role in mediating emotional arousal effects on memory consolidation.[26]

However, the noradrenaline system does not act alone in this respect. The glucocorticoids released by the adrenal cortex as a result of ACTH production enter the brain freely. There they have multiple effects through specific receptor systems. In particular, glucocorticoid effects on memory consolidation require them to act on the amygdala. Infusion of glucocorticoid agonists into the amygdala after training enhances retention whereas infusion of antagonists impairs retention. Again it can be concluded that the amygdala is the location for the impact of glucocorticoid enhancement on memory consolidation. Details of the neurotransmitter and neuroreceptor systems involved in the various pathways linked to the amygdala's role in memory consolidation are being steadily elucidated.[27]

A full account of the circuits involved in the total stress response will be very complex,[28] but it is clear that the amygdala is on one of the pathways leading to the initial readiness of the body to respond to danger signals. Subsequently, input from the body leads to noradrenaline and glucocorticoid activation of cells in the amygdala, and output from the activated cells has a considerable impact on the enhancement of memory consolidation by other brain structures. Furthermore, consolidation enhancement via the amygdala can be interrupted by the use of antagonists that act on the amygdala adrenoceptors. What then does this have to do with treatment of people suffering fom PTSD?

It has been found, first, that studies on humans produce similar results to those on animals:

> Humans with temporal lobe lesions show deficits in fear conditioning, as do individuals with lesions primarily confined to the amygdala ... Individuals with a rare disorder resulting in localised bilateral amygdala damage show the above deficits, and although they are able to understand logically that some situations are risky or will most likely have a negative outcome, they are largely unable to use this information to act accordingly.[29]

Functional neuroimaging shows that the amygdala is selectively activated when negative emotional stimuli are being processed and in fear conditioning. This and much other evidence supports the view that the amygdala plays a similar role in humans to that in animals in dealing with frightening situations.

Whilst the events are much more complex than in animals it can be considered that noradrenaline neurotransmission is therefore crucial in humans too because very traumatic events would lead to overproduction of the transmitter and thus overconsolidation of the memory of these events. As the traumatic memory causes the events to be relived in flashbacks and nightmares a positive feedback system could therefore lead to the further consolidation of the memory as the body responded again and again to the stress.

This idea of a direct relationship between noradrenaline and memory for emotional events has been tested in humans. Healthy subjects were either given a placebo or propranolol (which passes the blood brain barriers and antagonizes the action of noradrenaline) one hour before viewing a series of either neutral or emotionally stressful scenes. One week later people who had received the placebo had significantly better memories of the emotional slides but those who had received the propranolol did not remember them any better than the neutral ones.[30]

Such results have obviously led to efforts to prevent people developing PTSD, in one example giving victims of car crashes propranolol quickly after the event. Some observers, however, are concerned that such treatment might be used to enable people to carry out dreadful actions and retain no memory of them. Dr Leon Kass, chairman of the President's Council on Bioethics in the United States has been quoted as saying: 'It's the morning-after pill for just about anything that produces regret, pain, or guilt.' [31] A national co-ordinator for Vietnam Veterans Against the War agreed and argued that such treatment could 'make men and women do anything and think they can get away with it'. A different possibility, of course, is that those with malign intent might find means – through a chemical agent – to enhance PTSD, not prevent it.

We can initially conclude that the new systems biology brought about by the genomics revolution and associated scientific advances will undoubtedly open up complex behavioural systems not only to benign but also to malign manipulation. In considering the latter therefore, we have to ask just what might soon be open to manipulation.

2.2. Reassessing Cold War research

An initial approach to a more systematic investigation of the possibilities clearly is to look back at what was done in the Cold War period. It is important not to underestimate the scientific effort involved in that period. For example, a United States General Accounting Office report of 1994 noted:

> From 1952 to 1975, the Army conducted a classified medical research program to develop incapacitating agents ... Army documents identify a total of 7,120 Army and Air Force personnel who participated in these tests ...
>
> During the same period, the Army Chemical Corps contracted with various universities, state hospitals, and medical foundations to research the disruptive influences that psychochemical agents could have on combat troops. The Air Force also conducted

> experiments on the effects of LSD through contracts at five universities ...
>
> According to a CIA official, from 1953 to about 1964, the CIA conducted a series of experiments called MKULTRA to test vulnerabilities to behavior modification drugs.[32]

The net result of such research was suggested in the original SIPRI study which quoted a US Army manual of 1968 which argued that only two types of chemical agents aimed at the central nervous system were likely to be encountered in military use:

> *CNS depressants*. These are compounds which have the predominant effect of depressing or blocking the activity of the central nervous system, often by interfering with the transmission of information across synapses. An example of this type of agent is BZ ... Cannabinols and phenothiazine type compounds are other potential incapacitating agents which seem to act basically as CNS depressants. The primary effects of these agents, however, are to sedate and destroy motivation rather than to disrupt the ability to think.

On the other hand there are:

> *CNS stimulants*. These agents cause excessive nervous activity, often by 'boosting' or facilitating transmission of impulses which might otherwise be insufficient to cross synapses. The effect is to 'flood' the cortex and other higher regulatory centers with too much information, making concentration difficult and causing indecisiveness and inability to act in a sustained, purposeful manner. A well-know drug which appears to act in this manner is d-lysergic acid diethylamide [LSD]; similar effects are sometimes produced by large doses of the amphetamines.[33]

There is no mention of the disruption of peptide neurotransmission here but this is not surprising since understanding of this phenomenon would only come later.

However, by the 1990s there was definite military interest in the potential misuse of neuropeptides, as is evident from a 1990 US Army Intelligence Agency report on incapacitating agents research in

European communist countries. This called attention, for example, to a meeting in Moscow in the mid-1980s on 'Neuropeptides: From the Neuron to Behaviour'. It noted that many compounds of this type had recently been discovered and that they:

> have an extremely wide range of regulatory activity, including transmitting nerve impulses, control and release of large peptide hormones, opiate, and myotropic activity. A large number of these peptides are involved in the regulation of the complex process of behaviour and learning.

Later, the report also noted:

> Presently, it is known which neuropeptides are responsible for the sensations of hunger, satiety, or thirst, drowsiness, and elicit motor activity. When this information is combined with [that about] some of the synthetic neuropeptides which, in some cases, are a million times more active than the natural ones, the potential for a very strong behavior-modifying compound ... could exist.[34]

The review of the classified literature in the report is almost entirely blacked out in the available version, but the section ends 'though the open literature indicates that the Soviets have a strong interest in peptides, including looking at psychotropic effects which can induce fear and paralysis'.

The 1997 US Army textbook, *Medical Aspects of Chemical and Biological Warfare,* drawing on the work carried out beteween the early 1950s and early 1970s, suggests that virtually all psychochemicals can be classified into four groups: stimulants; depressants; psychedelics; and deliriants. It suggests that the drugs of interest pass the blood-brain barrier with ease and exert their dramatic effects on the functions of the central nervous system. It then continues:

> Their interference with higher functions (as opposed to basic vegetative functions, which are primarily under brainstem control) are of greatest relevance to the goal of producing military incapacitation. The higher functions of the brain (attention, orientation, perception, memory, motivation, conceptual thinking, planning and judgement) are more easily disrupted than are the more

> robust systems that regulate the physiological functions that are essential to life.[35]

Thus the textbook concludes that it is possible to disrupt these higher functions with lower amounts of agent than those that are required to produce lethal effects.

The text goes on to suggest that none of the stimulants such as amphetamines, cocaine, caffeine, nicotine, have sufficient potency to be of use as airborne incapacitants. Depressants such as barbiturates are similarly dismissed, and analogues of morphine are considered to have too small a gap between their incapacitating and lethal doses. However, the use of a fentanyl derivative of this type in the 2002 Moscow theatre siege demonstrates that this view was not universally shared.[36] Amongst the psychedelic compounds, most Western military interest during the Cold War period was in LSD, but although LSD is highly potent, it tends to produce unpredictable behaviour. This went against one of the key criteria set for an incapacitant since unpredictable behaviour could be extremely dangerous. The report suggests that interest focused on deliriants which produce 'an incapacitating syndrome, involving confusion, hallucinosis, disorganised speech and behaviour'. Many drugs could produce such effects, but anticholinergics such as BZ were regarded as the most likely to be used militarily. These drugs block the muscarinic effects of acetylcholine (neurotransmitter) in the central nervous system and BZ can produce delirium at low dose and with a high safety margin, according to the text.

That review of the Cold War period shows that muscarinic receptor systems in the brain could remain a target for new incapacitants. Beyond that, it suggests that the central mechanisms of interest are more likely to be higher brain functions. The newer neuropeptides and their roles could clearly be of interest to those with malign intent. However, we cannot make an adequate survey of the possibilities for malign misuse solely on the basis of what happened during the Cold War period.

We have already noted that a fentanyl derivative was used in Russia in 2002 to end the Moscow theatre hostage crisis. In the United States such agents and related opioids were still being researched in the early 1990s, as were agents that could affect α_2-adrenergic transmitter/receptor systems.[37] Also, some of the natural agents understood during the Cold War period were known to affect central nervous system processes – for example, fever pyrogens.[38] Furthermore, central nervous system processes related to circadian rhythms of sleep and alertness are increasingly being manipulated by the military – for example, so that pilots are fit to carry out extended missions.[39] Obviously, not all of these more automatic processes which are controlled by lower brain levels are inherently difficult and dangerous to affect by chemical means. And, of course, a great deal more is now known about the systems biology of circadian rhythms in mammals, thus opening them up to malign manipulation in many ways.[40] The point has already been made that the genomics revolution has opened up significantly more means of sophisticated manipulation of the central nervous system due to the vast amount of information it has made available on neuroreceptor sub-types. Though sometimes forgotten, the technology of drug delivery has also been revolutionized in recent decades as the pharmaceutical industry seeks more effective means of (benign) drug discovery. This necessarily also impacts on the possibilities for malign manipulation.[41]

Neuroscientists have always placed a strong emphais on understanding *systems*. As one new journal states in regard to its own scope:

> Neuroscience is the archetypal multidisciplinary science encompassing a variety of fields that share the common goal of attempting to provide a complete understanding of the structure and function of the central nervous system. A series of advances in molecular, developmental and cognitive neuroscience ... *have now rendered some of the most enduring neurobiological questions increasingly tractable.* (emphasis added)[42]

It is therefore no surprise to see neurobiology examples being cited in current reviews of the new systems biology[43] or to see extensive use being made of model organisms such as *Caenorhabditis elegans* whose molecular genetics has been elucidated in the discipline of neurobiology.[44] These facts argue for keeping a very open mind as to the possibility of advances that might be misused. A final reason for keeping an open mind and a broad approach, of course, is the scope and pace of change in the biological sciences in general at the present time. This point was made strongly in the United Kingdom background paper on relevant scientific and technological developments for the

Fifth Review Conference of the Biological and Toxin Weapons Convention which stated:

> Throughout the various studies and consultations carried out by the UK to inform this review, it has been clear that the rate of change in science and technology fields relevant to the BTWC has been much greater than in the previous five year period, that is between the third and fourth Review Conferences.[45]

What follows is therefore best considered illustrative rather than in any way definitive of the problems we may face.

2.3. The central nervous system

Before investigating two more detailed examples of potential misuse, it is necessary first to describe something of the structure of the human nervous system. This is divided into the central nervous system (brain and spinal cord) and the peripheral nervous system.[46] Information from peripheral sense organs is received via afferent pathways and processed within the central nervous system. Output from the central nervous system is sent via efferent pathways to the somatic nervous system (muscles) and to the autonomic nervous system (heart, gut, glands, etc.).

The human body is organized bilaterally, with two arms, legs, eyes and so on. It is also organized segmentally, so that in the vertebral column the basic input and output nerve pathways are repeated in each segment. The brain, then, is an expansion of that basic segmental organization associated with the massive sensory input at the front end of a moving living organism.[47]

The most useful way to understand this massively complex structure is by reference to its growth during development. As the embryo develops, there are at first three primary brain regions: the prosencephalon (forebrain); the mesencephalon (midbrain); and the rhombencephalon (hindbrain). Within a few weeks the forebrain and hindbrain each divide in two. The forebrain gives rise to the telencephalon and diencephalon while the hindbrain gives rise to the metencephalon and the myelencephalon. The telencephalon of the forebrain then develops into the cerebrum with its hugely expanded, characteristic, cerebral hemispheres which cover the top and side surfaces of the brain (Table 5.1). The surface of the cerebrum is made up of central nerve

Table 5.1: Some structures of the brain

FOREBRAIN
Telencephalon
Cerebral cortex
— Archicortex
— Hippocampus
Basal ganglia
— Amygdala
— Striatum
Diencephalon
Hypothalamus
Thalamus
MIDBRAIN (Mesencephalon)
Tectum
HINDBRAIN
Metencephalon
Cerebellum
Pons
Myelencephalon
Medulla oblongata

Source: From Dubin, *How the Brain Works*, note 48.

cells in areas such as the primary motor and somatosensory cortex regions. Other central nerve cells are are grouped in deeper structures often called nucleii or ganglia. There are also large regions of connecting nerve fibres throughout the brain (termed white matter because of the colour of the sheathing around the nerve fibres). In general, it can be said that the parts of the brain nearer the spinal cord deal with more automatic functions (such as heart and temperature control) and that higher functions are located more in the forebrain.[48] Some of the larger structures in the brain which are mentioned later in this chapter are listed in Table 5.1.

2.4. Brain cholinergic systems

The central nervous system is made up of billions of individual nerve cells (neurons). Transmission of information *within* a neuron is by electrical means and the transmission of such nerve impulses can be recorded with suitable equipment. However, most transmission of information *between* neurons or between neurons and effector systems (muscles, for example) is by chemical means. The first chemical neurotransmitter to be discovered was acetylcholine, a small molecule. Subsequently, many other small molecule neurotransmitters, such as noradrenaline and dopamine, have been discovered. More recently, it has been discovered that various neuropeptides (molecules made up of short strings of amino acids) also function as neurotransmitters and that sometimes a particular neuron can employ both a small molecule neurotransmitter and a neuropeptide at the same time.

What is of interest here is that the nerve agents weaponized as lethal agents in past offensive programmes interfered with acetylcholine's functions in what are termed *cholinergic* transmission systems. It has long been known that there are two distinct types of cholinergic synapse, a synapse being a junction between presynaptic and postsynaptic nerve cells where information is transferred. At one of these types of synapse the effect of acetylcholine can be mimicked by nicotine and at the other it can be mimicked by muscarine, a chemical extracted from a mushroom. So-called nicotinic synapses are found, for example, at neuromuscular junctions and necessarily operate rapidly. So-called muscarinic synapses found, for example, in the brain operate more slowly and in these the post-synaptic receptors for acetylcholine are of the G-protein-coupled family which has been discovered over the last decade. Despite such differences, the acetylcholine produced at all cholinergic synapses is destroyed by an acetylcholinesterase enzyme so that its effects are not prolonged. The weaponized nerve agents were found to bind to the acetylcholinesterase and thus to inhibit its action. Consequently, cholinergic synapses were flooded with acetylcholine and death from the consequent malfunction of essential body systems quickly followed.[49]

The incapacitating agent BZ, which was also weaponized by the United States in the 1960s, interferes with the operation of acetylcholine at muscarinic synapses in a different way. BZ locks onto the receptors of the postsynaptic cell in this type of synapse and so prevents acetylcholine exerting its normal effects.[50] As most of the cholinergic synapses in the brain are of the muscarinic type, it is not surprising that BZ was found to have severe effects on behaviour (Table 5.2). It does not appear, however, that BZ was ever used in warfare. This is hardly surprising because the effects were too wide-ranging to be predictable in any particular individual. The question here is whether advances in neuroscience have produced a situation in which more controllable changes in behaviour could be produced.

Table 5.2: Effects of BZ on human beings

Rapid pulse
Dry mouth and blurred vision
Poor coordination
Restless activity
Stupor
Confusion, incoherence, hallucinations, disorientation
Irritable, suspicious and uncooperative
Inability to solve problems or remember information

Source: From Dando, *A New Form of Warfare*, note 46.

Cholinergic neurons of the brain are of two types. There are interneurons which are intrinsic to the striatum of the basal ganglia (that is entirely confined to a group of neurons of the forebrain), and projecting neurons which have their cell bodies in one part of the brain but then have axons that deliver their output in a different part.[51] There are a number of different pathways of projecting neurons some of which are known to be crucial in important diseases. For example, the neurons of the nucleus basalis of Maynert project extensively in the cerebral cortex, and it is these neurons which are known to degenerate in Alzheimer's disease.

Nicotinic receptors act by the transmitter simply opening a pore in the post-synaptic neuron membrane and thus allowing a flow of ions which changes the electrical characteristics of that cell. Muscarinic receptors operate differently. These act via various G-proteins to change both the electrical and more complex metabolic activities of the post-synaptic cell. To date, five different muscarinic receptor subtypes have been discovered:

> The muscarinic family has five known members designated M_1–M_5 based on the order in which they were cloned ... The M_1, M_3 and M_5 *stimulatory* receptors couple primarily to the mobilisation of intracellular calcium ... The M_2 and M_4 *inhibitory* receptors negatively modulate adenylate cyclase to reduce cytoplasmic levels of cAMP. (emphasis added)[52]

In other words, the odd-numbered receptors excite the post-synaptic cell whilst the even-numbered muscarinic receptors decrease its electrical activity.

The structure of muscarinic receptors has been strongly conserved across species and sub-types during evolution and it has therefore been difficult to find potent agonists/antagonists for one sub-type at a time in order to elucidate their different functions.[53] However, as the revolution in the life sciences has progressed, it has become possible to breed different strains of mice each lacking one of the different sub-receptor types (so-called 'knockout mice') and to discover where the different sub-types are to be found in the brain and, increasingly, what their different functions are in these different locations.[54] The inhibitory receptors may be located on the presynaptic neuron and function there as inhibitory autoreceptors. Therefore when acetylcholine is released by the presynaptic neuron where such receptors occur there is an inhibitory feedback loop which serves to limit its production. A further study of the physiology of neurons in brain regions of knockout mice concluded:

> autoinhibition of ACh release is mediated primarily by the M_2 receptor in hippocampus and cerebral cortex, but predominantly by the M_4 receptor in the striatum. These results, together with additional receptor localization studies, support the novel concept that autoinhibition of ACh release involves different mAChRs [muscarinic acetylcholine receptors] in different regions of the brain.[55]

Such results are obviously of great interest to those trying to help people with Alzheimer's disease.

If a selective agonist could be found for the M_1 receptor it might be possible to increase the level of excitation in the cortex and make up for the deficiency in acetylcholine, but this approach has not proved to be successful. Another possibility is to find a selective antagonist to the M_2 receptor and thus to block the inhibition of the presynaptic cell and increase the production of acetylcholine back to functional levels. Several drug companies are pursuing this strategy. As the Schering-Plough Research Institute of Kenilworth, New Jersey reported:

> SCH211803 is a functional M_2 antagonist ... The compound increases ACh release from hippocampus and striatum of conscious rats.

Furthermore, the report continued:

> Improvements in learning and memory tasks with SCH211803 have been demonstrated in rats using the Morris Water Maze and the young rat passive avoidance response test. SCH211803 also improves performance in an operant model of cognition in squirrel monkeys.[56]

The results thus confirm that selective blocking of M_2 receptors increases acetylcholine release and this causes improvements in cognition – just what might help Alzheimer's sufferers. More recent work by this same group has been devoted to improving the oral efficacy of the M_2 receptor antagonist by chemical modification of the molecular weight and lipophilicity of the lead compounds.[57] Clearly, however, it could also be possible to find a chemical which acted in such a way as to be more effective agonist at M_2 receptors than the natural transmitter and thus to increase the negative feedback loop. Such a chemical might therefore provide a specific disruption of function instead of the multiple and unpredictable disruption caused by BZ.

2.5. Narcolepsy

So far in this chapter we have discussed something of the neurobiology of awareness, fear and cognition. But these higher functions rest on a whole set of more automatic homeostatic functions. No animal could operate without effective regulation of, for example, temperature or blood pressure. These kinds of functions are normally regulated from centres in lower parts of the brain near the junction with the spinal cord. Here we will consider an aspect of one regulatory system – sleep – and, in particular, one of its malfunctions, *narcolepsy*. Before discussing narcolepsy and how its investigation will likely lead to means of helping sufferers but also open up new roads to misuse, a brief review of modern knowledge of biological clocks will be necessary.

Many of our basic physiological functions exhibit a circadian (daily) rhythm. Most noticeably, we tend to sleep each night for about eight hours, but other functions – core temperature and production of pituitary hormones, for example – also exhibit such a rhythm. If sensory cues, most importantly light, are eliminated then our sleeping/waking cycle will elongate from 24 to about 30 hours. It is therefore clear that sensory inputs affect the basic circadian cycle, but what has been dramatically demonstrated recently is that the basic rhythm is driven by an internal clock located in cells of the superchiasmatic nucleus (SCN) of the anterior hypothalamus. The output from this intrinsic clock then flows to complex circuits in other parts of the hypothalamus to regulate the various circadian cycles. Light input direct from the retina via the retino-hypothalamic tract (RHT) synchronizes the output of the SCN with the 24-hour cycle.[58] From the systems biology viewpoint, what is particularly important is not only that the genetic basis for the sinusoidal form of output from the SCN (with neuronal firing peaking during the day) has been elucidated, but how this output is integrated from the single cell through the SCN nucleus, the brain and then the behaving animal is also increasingly being understood.[59] From the neurobiologists' viewpoint, it is also crucial to note that the role of the different neurotransmitters in circuits governing the various physiological functions is also being steadily clarified.[60]

Sleep, of course, is not just a quiescent state the opposite of wakefulness. During the second half of the last century a great deal was learned about what happens when we sleep from recordings of the electrical activity of the brain picked up from electrodes placed on the scalps of volunteers. When we are awake these electroencephalography (EEG) recordings are of low amplitude and high frequency. When we fall asleep we pass through four phases of what is called slow wave sleep in which the EEG recordings have high amplitude and low frequency. If awoken from such sleep, people are confused, find it difficult to think clearly and easily go back to sleep. However, at about 90-minute intervals a quite different type of sleep appears. This type of sleep is called rapid eye movement sleep (REM), or paradoxical sleep because the EEG resembles that of the awake state. In this kind of sleep people dream and muscle tone is absent apart from the extraocular eye muscles producing rapid eye movements.[61] Again, the mechanisms underlying this behaviour are being elucidated – even if we still cannot explain *why* we sleep.

Despite much effort to find cures, there are many people who suffer from sleep disorders such as insomnia, obstructive sleep apnoea, and narcolepsy. So there is every good reason for further investigation of the underlying neuronal mechanisms.[62] Narcolepsy is characterized by four essential features:

> excessive daytime sleepiness (EDS), catalepsy (sudden loss of muscle tone in response to strong emotion such as laughter or anger,

> hypnagogic hallucinations (dream-like experiences occurring at sleep onset), and sleep paralysis (the inability to move while falling asleep or upon waking).

The total amount of sleep and REM sleep is of the same order as in people without narcolepsy but, clearly, the control mechanism is severely disrupted with two main problems:

> first, an inability to maintain wakefulness, and second, intrusion of REM sleep into wakefulness or at sleep onset resulting in hallucinations, sleep paralysis, and possibly cataplexy.[63]

Despite this condition being quite widespread and debilitating, until recently very little was known about its causation. It was known that some dog families exhibited very similar symptoms to those of human narcolepsy, and this suggested a genetic basis for the disease. However, as there was also a strong link with aspects of the human leukocyte antigen (HLA) system, an autoimmune disorder was also strongly suspected and this, of course, might have an environmental trigger.

The whole field of research can reasonably be described as being revolutionized over a few years at the turn of the century through the discovery of two neuropeptide transmitters produced by cells of the hypothalamus. These hypocretins (Hcrt-1 and Hcrt-2), which are also known as orexins, are clearly the key to understanding narcolepsy, and a good deal of the normal sleep mechanism. The familial canine narcolepsy cases are associated with genetic mutations of this system, mice with targeted deletions of the gene for the precursor of these peptides display symptoms of narcolepsy, and the majority of humans with narcolepsy and the associated HLA characteristics have no detectable hypocretins in their cerebrospinal fluid.

The hypocretins were discovered in 1998. First, a paper describing a search for the most abundantly expressed mRNAs exclusive to the rat hypothalamus allowed the determination of a prepropeptide with cleavage sites suggesting two peptides with a weak structural resemblance to secretin, a member of the incretin superfamily of peptides – hence 'hypocretins' from the hypothalamus.[64] Then a second paper was published in the same year describing the search for endogenous ligands for 'orphaned' G protein-coupled receptors (GPCRs).[65] Cultured cells expressing the receptor on their surface were challenged with purified extracts of hypothalamic tissues and cellular activity changes were monitored. This allowed the isolation of peptides orexin A and orexin B (now known to be equivalent to Hcrt-1 and Hcrt-2). The name 'orexin' was selected because the original tissue was located near the hypothalamic feeding centre, injection of the peptides into the brain increased food intake and the peptide precursor was upregulated in fasting. It is now known that the hypocretin cells project widely in the brain and are involved in multiple functions. The hypocretin nomenclature is used because it was that of the first publication. Progress in elucidating the nature of the narcolepsy has been phenomenal since these discoveries.

Human Hcrt-1 is a 33-amino-acid-chain peptide and is identical to the mouse, rat, bovine and porcine peptides. Hcrt-2 in humans is a 28-amino-acid chain which has two different substitutions compared with rodent Hcrt-2 and one compared with porcine and canine Hcrt-2s. Clearly, these peptides are strongly conserved, suggesting important functions. Both peptides have similar affinities for the hcrtr2 receptor but Hcrt-1 has greater affinity than Hcrt-2 for the human hcrtr1 receptor. In all experiments carried out so far hypocretins have had excitatory effects on post-synaptic cells. The noradrenergic locus coeruleus neurons discussed previously are densely packed with hcrtr1 receptors, but not hcrtr2 receptors.

Narcolepsy affects 20–60 people per 100,000 of the population in Western countries. This is about the same level of incidence as Parkinson's disease or multiple sclerosis, but unlike those diseases it usually begins in the teens or twenties when it is very debilitating at a crucial formative period and continues to be so for many years. At present most patients require drug treatment:

> EDS [excessive daytime sleepiness], the most disabling symptom, is treated with amphetamine – like stimulants or modafinil. These compounds act by stimulating dopamine release ... and/or inhibiting dopamine reuptake (modafinil).[66]

Other symptoms have to be treated with other drugs and none are free of side-effects. The need to find better treatments is obvious and research will clearly continue to achieve this end.

Work on dogs with narcolepsy has supported the view that a reciprocal aminergic-cholinergic mechanism is involved. High activity of cells producing these neurotransmitters leads to wakefulness and characteristic normal wakeful EEG activity. In normal (NREM) sleep, as the characteristic synchronization of the EEG occurs, monoamine and cholinergic activity decreases. During REM sleep there is little monoaminergic activity but cholinergic systems are active. Clearly, drugs that affect these systems can have profound effects, as can natural agents such as the hypocretins which strongly innervate the noradrenergic neurons of the locus coeruleus (LC) and have excitatory effects. In particular, human beings with narcolepsy have low or non-existent levels of hypocretins in their brains and thus would lack this excitatory input to the locus coeruleus noradrenergic nrurones. These would therefore be much less active, which probably explains many of the symptoms such as excessive daytime sleepiness. Certainly, direct application of HCRT-1 onto cells of the locus coeruleus leads to an increase in wakefulness and a decrease in sleep in rats.[67]

In some dog families narcolepsy is caused by a gene mutation. In humans there may be a genetic susceptibility in some people, but an autoimmune causation – presumably with an environmental trigger involved – is the most likely explanation for most human cases. If this is indeed found to be the correct explanation, given the ongoing elucidation of the mechanisms of normal sleep patterns and the abnormal sleep patterns of narcolepsy, it is not impossible that means will be found, by those with malign intent, to trigger narcolepsy. Such a disruption of normal functioning would, of course, be profoundly debilitating for an individual, or groups of people, affected.

3. Conclusion

It is known that during the Cold War both sides tried to find means of interfering with the operation of the human central nervous system. Given the state of scientific knowledge at the time, these efforts did not produce agents more sophisticated than the lethal chemical nerve agents. The examples given here strongly suggest that if the situation has not already changed, it is very likely to do so in the near future. We face the strong possibility that agents developed for malign/hostile purposes will be perfected in order to achieve precise effects on human behaviour.

6
Double Assault: Malign Manipulation of the Neuroendocrine-Immune System

1. Introduction

Concerns about biological weapons and biological terrorism have increased over the last decade and, particularly since the events of 11 September 2001 in the United States. There has been a growing belief that large-scale biological weapons attacks are becoming more likely.[1] The medical profession has been amongst those groups which have devoted more and more attention to what might need to be done in the event of an attack.

If standard accounts of the effects of well-known biological weapons agents are reviewed – like those of agents on the Centres for Disease Control (CDC) Category A list such as smallpox, anthrax, plague, botulism, tularemia and viral haemorrhagic fevers – it becomes very apparent that easy diagnosis of the cause of any such attack would be far from straightforward.[2] If people had fallen victim to any one of a number of different biological weapons agents on the Category A list, they would often present with the same 'flu-like' symptoms. This would also be true for some of the lesser agents on the Category B and C lists.

Not surprisingly, medical specialists have begun to consider how their particular expertise might be best used to assist in patient diagnosis and better care. It has been argued, for example, that:

> Many of these agents (such as anthrax) may present with headache, meningitis, or mental status changes in addition to

> fever and other symptoms and signs … *Thus, a neurologist may be consulted acutely to aid in diagnosis.*[3] (emphasis added)

Similarly, it has been noted that early recognition of the characteristic clinical signs is important for rapid initiation of therapy and minimization of casualties, and that:[4]

> Neurophysiological investigations when integrated with clinical features are helpful in early identification of some of these agents.

So it is clear that neuroscience is one of the specialisms that can be recruited to help deal with a biological weapons attack should such an event occur.

We can see this clearly, for instance, in regard to the possible use of deadly botulinal toxin. It is well known that this toxin disrupts the operation of parts of the nervous system by interfering with the normal synaptic release of the key neurotransmitter chemical acetylcholine. The clinical features of botulism are very similar, regardless of how the toxin gets into the body:

> Botulism preferentially affects the cranial nerves, causing … poorly reactive dilated pupils, disconjugated gaze, facial dipleglia … Generalized weakness and respiratory compromise develop.[5]

As would be expected from the interruption of acetylcholine neuromuscular transmission, electrophysiological studies show smaller than normal motor unit potentials in the victim's muscles. The only specific treatment available is passive immunization with an antitoxin. This will not reverse the paralysis but can help to stabilize the decline. The antitoxin available is from an equine source and thus there can be complications for the patient such as anaphylaxis. The main therapy is supportive care – often involving maintenance of ventilation – over the longer term until normal functions recover.

Perhaps surprising for those outside the medical profession is the fact that all three forms of anthrax infection – inhalation, cutaneous and gastrointestinal – may be complicated by meningitis. An anthrax biological weapons attack would most likely be an attempt to spread the spores on the air to cause inhalation anthrax. Once inside the body, the spores begin to grow out into bacteria, in particular in some lymph nodes, and the bacteria produce damaging toxins. In most victims symptoms would be expected within a week. Initial symptoms would include:

> fever, chills, myalgia, cough and sore throat. Substernal chest pains, dyspnea, abdominal pain, nausea, and vomiting are common.[6]

However, of the ten inhalation anthrax victims of the attacks in the USA in the autumn of 2001, eight also had neurological problems such as 'headache, confusion, blurred vision, visual field distortions'.[7]

Fortunately, anthrax meningitis can be treated with a multi-drug antibacterial regime. However, although there is an anthrax vaccine, it requires multiple doses over a protracted period and annual boosters and is also in relatively short supply. Vaccination is therefore not really an option for mass protection at present.

The connection between neuroscience and possible biological attacks with classical agents may not be readily apparent to the general public, but the connection with immunology is perhaps less surprising. The body's immune system is its defence mechanism, so when the victim of an attack begins to be affected by an invading microorganism it is to be expected that the defence mechanism will be activated. What may still be a surprise is that microorganisms have evolved means by which to invade or even to attack the defensive immune system itself.

As we have seen in Chapter 4, the immune system comprises the ancient, non-specific, innate immune system and the more recently evolved specific, adaptive immune system. The ancient innate immune system is vital both for providing an initial rapid response to infection, and also in activating the adaptive system.[8] A crucial part of the early innate immune system response is provided by two proteins of the complement system called C3b and C4b. These proteins are vital in the recognition and destruction of invading microorganisms.

However, large viruses such as smallpox (and other pox viruses) have evolved ways of disrupting the immune system. It has been found, for example, that in vaccinia (the viral strain used to make smallpox vaccine) there is a vaccinia virus complement control protein (VCP) which binds to and inhibits C3b and C4b. In 2002, by examining genomic sequence databases, scientists discovered that there was an analogous protein in pathogenic smallpox. This protein was named smallpox inhibitor of complement enzymes (SPICE). It was found that:

> SPICE is nearly 100-fold more potent than VCP at inactivating human C3b and six-fold more potent at inactivating C4b.[9]

Thus the greater activity of SPICE in inhibiting C3b and C4b could be one of the crucial reasons for the greatly increased pathogenicity of smallpox over the vaccinia strain.

The interactions between potential biological weapons agents and the immune and nervous systems can be much more complex even than that of the smallpox inhibitory protein example. This is well illustrated by the case of the Category B – but previously weaponized – incapacitating staphylococcal enterotoxin B (SEB).[10] Although this was probably not clear when the toxin was weaponized by the United States in its early Cold War offensive biological weapons programme, SEB exerts its incapacitating effects through a particular action on the adaptive immune system. The toxin was particularly attractive to weaponeers because of the very low dose required to incapacitate:

> The dose that is incapacitating for 50% of the human population exposed (also called the effective dose [ED_{50}]) was found to be 0.0004μg/kg, and the dose that is lethal for 50% of the human population exposed (LD_{50}) was estimated to be approximately 0.02μg/kg, both by the inhalation route.[11]

The effective dose, though much lower than that required for synthetic chemicals, was additionally much smaller than the lethal dose. This fact made the toxin more potentially usable for incapacitation. What is of interest here is the mechanism by which the toxin subverts the immune system.

SEB is, in fact, one of at least seven distinct enterotoxins produced by *Staphylococcus aureus* bacterial strains. When ingested, SEB causes classical symptoms of food poisoning 'violent vomiting, diarrhea, fever (41°C), and in severe cases, lethal shock'.[12] The symptoms set in within six hours of ingestion, but although unpleasant, are usually self-limiting within 48 hours. Lethal doses of the toxin administered by inhalation to monkeys caused similar symptoms, but the animals died some 50 hours after receiving the toxin. The cause of death was found to be severe pulmonary oedema. SEB and associated toxins are known technically as 'super-antigens' because they overstimulate the adaptive immune system.[13] The details of this mechanism need not concern us here, but the result of the SEB binding is that there is a massive release of cytokines (chemical messengers of the immune system) which produces the incapacitating (or lethal) effects on the victim's body.

The impact of administration of the toxin, however, is not confined to the immune system. It has recently been demonstrated that there are also quite significant changes induced in the endocrine (hormonal) and nervous systems and even behavioural changes. It has been shown that administration of SEB activates the hypothalamus-pituitary-adrenal (HPA) stress response axis,[14] and that the challenge with SEB leads to the new behaviour of reduced consumption of mainly new foods or of food given in a new environment.[15] The situation is even more complex because, as one recent review noted:

> Thoughts, feelings, and behaviours affect our health and well-being. Recognition of the importance of these influences on health and disease is consistent with evolving conceptions of mind and body and represent a significant change in medicine and the life sciences.[16]

In a full account of the relevant interactions it would therefore also be necessary to assess how psychological factors affect, and are affected by, processes in the nervous, endocrine and immune systems. Here, however, we will concentrate attention on the interactions between the nervous and immune systems.

The chapter begins with a brief account of how these systems are interconnected and then two major sections discuss how the nervous system affects the immune system and vice versa. Particular attention is given to discussion of the means by which these complex interactions might be disrupted by those intent on causing harm.

2. Neural regulation of the immune system

In order to survive, living organisms have to maintain a relatively stable internal environment. This stability is threatened by both internal factors, such as a disease process, and external factors such as stress. Human and animal bodies have therefore evolved complex regulatory (homeostatic) mechanisms which counteract such disruptive factors and re-establish the internal stable state.

Of interest here is how the nervous system responds to *stress*, as was briefly described in Chapter 5. The central nervous system can impact on the immune system in two ways, via:

> (a) the hormonal stress response and the production of glucocorticoids, and (b) the autonomic nervous system with the release of noradrenaline.[17]

The central nervous system can also affect the immune system through the peripheral release of neuropeptides but that issue will not be addressed here. What is of particular interest is the brain's response to stressors through the hypothalamus producing corticotropin-releasing hormone (CRH, also referred to as corticotrophin-releasing factor or CRF), which stimulates the pituitary gland to produce adrenocorticotropic hormone (ACTH), which in turn causes the adrenal gland to secrete the immunosuppressant glucocorticoids. Whilst the mechanism is clearly complex, with many other feedback loops being involved,[18] this hypothalamus-pituitary-adrenal (HPA) axis is the essential element for our purposes here. The total HPA general adaptive response can also be activated by immune cytokines (such as interleukins) produced in reaction to pathogens. Then, as in response to stress, the glucocorticoids eventually generated inhibit the further production of such cytokines.

There is now an accumulation of evidence which shows that deregulation of this system can cause a variety of diseases. As a background paper from the US National Institutes of Health summarized recently:

> Ideally, stress hormones damp down an immune response that has run its course. When the HPA axis is continually running at a high level, however, the damping down can have a downside, leading to decreased ability to release the interleukins and fight infection. [...] Conversely, there is evidence that a depressed HPA axis, resulting in too little corticosteroid, can lead to a hyperactive immune system and increased risk of developing autoimmune diseases – diseases in which the immune system attacks the body's own cells.[19]

This evidence comes from a very wide range of sources. At one level the impact of the nervous system on the function of the immune system can be demonstrated in animal experimentation, for relatively simple processes. For example, subjecting animals to quite modest levels of stress can greatly affect their ability to fight bacterial infection of cutaneous injuries. In one such study mice were subjected to restraint stress by placing them in loosely fitting, well-ventilated tubes overnight.[20] The function of the immune system in these restraint stress (RST) mice was then compared with that of control mice which were food and water (FWD) deprived for the same period but allowed to roam freely around their cages. After three days of such treatment female mice from both groups were anaesthetized and in each a standard wound made between its shoulder blades (where it could not be reached by the mouse). Restraint or deprivation was then continued for a further five days for each group.

The *Streptococcus* bacterium is a normal infection of the particular mouse strain used in the experiment so some animals in each group were also inoculated (RST-IN or FWD-IN) with a standard amount of these bacteria at the time of wounding. It was found that RST mice had a 30 per cent delay in wound healing at days 3 and 5 after wounding compared with the control FWD mice. Wound healing was also accelerated in the FWD-IN (inoculated mice) but not in the RST-IN mice as compared with the FWD and RST non-inoculated mice respectively. After 13 days the number of bacteria was greater in the RST-IN mice as compared with the FWD-IN mice and there were also many more opportunistic invasive bacterial species. There was clearly a difference in immune function in the RST and FWD mice. It was further found that there was a nearly-threefold increase in serum corticosterone levels in the RST mice as compared with the FWD mice and treatment of RST mice with a glucocorticoid receptor antagonist to prevent the damping down of the immune system produced a significant reduction in bacterial numbers in the RST mice. The authors of the paper therefore conclude quite reasonably that their results show that 'disruption of homeostasis by stress can significantly impair the ability to control and eradicate bacterial infection during wound healing'.

At a different level, there is much evidence of the impact of the nervous system on the immune system in humans.[21] Even short-term stress can have an effect. For example, students sitting examinations were found to have a significantly slower (40 per cent) rate of healing of a wound on the hard palate if the wound was made three days before an examination than if it was made in the same individuals during the summer vacation. Interleukin-1 levels – an important indicator of immune function – were also substantially lower during the examination period.

Not surprisingly, greatly elevated stress, such as from having to care for a relative who is ill or from having experienced a major disaster such as an earthquake or hurricane, has also been found to cause more pronounced dysregulation of the nervous system and thus the immune system. Furthermore, it has been found that social support and positive personal relationships can mitigate the effects of such stressors whereas negative relationships or prolonged thinking about the trauma or disaster can make the effects worse.

Though it is difficult to pin down the precise mechanisms involved, one recent review noted that:

> A growing literature supports the hypothesis that psychosocial factors have clinically signficant relationships with immune-related health outcomes, including infectious disease, cancer, wound healing, autoimmune disease, and HIV.[22]

For example, in one study people who reported long-lasting interpersonal difficulties were much more likely to develop a cold following inoculation with a virus than those with happier social relationships.

Stress is also involved in the modulation of the immune responses through mechanisms involving the neurotransmitter serotonin. This is a biologically active amine that plays a prominent role in the regulation of processes such as mood, appetite and sleep. During neurotransmission, serotonin is released from neurons into the synaptic cleft between cells. The amount of serotonin available for neurotransmission in the synaptic cleft is regulated largely by the serotonin transporter involved in the reuptake of serotonin after it has been released.[23]

In this regard it is also becoming clear that there are individual differences at the genetic level which can have a major bearing on response to stress. One recent study reported on 847 white New Zealanders who had been tracked from birth in the early 1970s through to adulthood.[24] It had been found that there are two forms of the gene for the serotonin transporter. The two versions are labelled *short* and *long* and of the people studied 17 per cent had two copies of the short version, 31 per cent had two copies of the long version and 51 per cent had one copy of each. Startlingly, the researchers found that having one or two copies of the short version had a pronounced behavioural impact:

> Among people who suffered multiple stressful life events over 5 years, 43 percent with one [the short] version of the gene developed depression, compared to only 17 percent with [the long]... version of the gene.[25]

It was also found that people with only the long version of the gene were no more subject to depression – no matter how many stressful events they experienced – than those totally spared stressful events.

The gene involved has a slight variation in a region which acts to control the switching on and off of transporter protein production. The short variant makes less protein and therefore there is a longer binding and function of the serotonin neurotransmitter before it is cleared from the synapse. Significantly, hypothalamic corticotropin-releasing hormone neurones are known to receive a positive input from serotonin neurones.[26] Severe depression, of course, resembles a chronic stress response with production of high glucocorticoid levels possibly being responsible for lower immune system functions.[27]

2.1. Malign manipulation?

The question then is how might such an increased level of understanding of these interactions be misused by those intent on causing harm? Given the complexity of the interactions, there could be many possibilities. One obvious possibility which stands out in the literature is the link between the disruption of metabolism and frailty in old age.

Frailty is not a simple characteristic to define, but it clearly includes factors such as loss of muscle strength and weakness, limited mobility, being underweight and failing to use nutrients effectively.[28] Eventually, the progressive decline means that many elderly people cannot cope alone and require assistance. Some older people, nevertheless, are relatively unaffected and continue to be able and active. So there must be differences between those who manifest major signs of frailty and those who do not, and there appears to be a strong link beween high levels of circulating interleukin 6 (IL-6) and frailty. IL-6 plays a role in many diseases that are prevalent in the elderly. As one report noted:

> These include diseases as different as coronary heart disease, stroke, congestive heart failure, osteoporosis, arthritis, depression and dementia.[29]

These diseases are major contributors to disability in older people. The report tested the link between high levels of serum IL-6 and disability in a large population in the USA. Over a thousand older people (65 or older) who had no disability consented to having a blood sample taken and were then interviewed four years later. The 283 people who had developed disability were then compared with 350 selected at random from those who had not. The results of the study were clear-cut:

> Using data from a large population-based prospective study, we found that older persons who are completely independent in ADLs [activities of daily living] and mobility and have circulation levels of IL-6 greater than 2.5 pg/mL are at risk of functional decline over the subsequent four years.[30]

The authors went on to suggest that IL-6 could have these effects through its known role in muscle wasting or in inflammatory processes but that in general '[t]hese data suggest that IL-6 is a global marker of impending deterioration in health status in older persons'.[31] They also suggested that as the high IL-6 levels result from dysregulation of normal physiological mechanisms, it may be possible to find means to reverse the process medically.

Not surprisingly, given such findings, IL-6 is not normally detected in the serum of healthy young individuals unless there is trauma, infection or stress. Then, IL-6 is expressed and contributes to typical inflammatory processes. It follows that the IL-6 gene must be under tight regulatory control, for example from secondary sex steroids – the decline of which in older people may account to some extent for their higher levels of circulating IL-6.[32] If IL-6 levels are increased in healthy individuals it is very likely that this would have severe effects. A study of young and middle-aged rhesus monkeys which received low doses of IL-6 daily for a month showed that: '[t]hese animals lost 10% of their body weight ... became osteoporic and anemic'.[33] Thus an obvious route for malign manipulation would be to find a means of increasing IL-6 levels in healthy people.

The human IL-6 gene is located on chromosome 7p21 and human IL-6 has 212 amino-acids with a signal sequence of 28 amino-acids. The mouse gene for IL-6 is on chromosome 5, consists of 211 amino-acids and has a signal sequence of 24 amino-acids. There appears to be a variety of regulatory mechanisms for expression of the gene and these are conserved between mouse and human which 'has been taken to indicate the importance of IL-6 gene regulation'.[34] From our point of view here it is interesting to note that glucocorticoids have been shown to inhibit IL-6 production in a range of different tissue types. The gene is regulated via a number of routes including corticosteroid interaction with the glucocorticoid receptor (GR) leading to the binding of the IL-6 promoter and thus to inhibition of the gene. Theoretically at least, one could consider the possibility of an irreversible blockage of the GR preventing such inhibition and leading to continuous overproduction of IL-6 with the consequent disablement of those affected – even the young and healthy.

Just how susceptible the immune system is to modulation through the nervous system can be seen in a recent study on the effects of low doses of the chemical weapon nerve agent sarin.[35] Sarin is a potent organophosphorus nerve agent that irreversibly inactivates acetylcholinesterase activity. This is an enzyme that is needed to degrade the neurotransmitter acetylcholine after it binds to receptors on muscle fibres, causing the muscle to contract, that is, to do work. Degredation of acetylcholine allows the receptors and therefore the muscle cells to return to their original state of relaxation after contraction. If acetylcholinesterase is inactivated, the neurotransmitter continues to stimulate the muscle to contract, until it is completely exhausted and cannot perform work. Sarin can cause seizures and even death in high doses. In their study on low dose effects, Kalra et al.[36] found that rats exposed to inhalation of even low, subclinical doses of sarin (0.4 mg/m^3 for 1 hour per day for 5 days) showed significantly reduced antibody (three-fold) and T lymphocyte-mediated (2-fold) responses compared to those animals that had not been exposed to the nerve agent at all.

2.2. Molecular mechanisms

At the beginning of this chapter we referred to the work of the US National Institutes of Health. In a summary of the work of the NIH section on Neuroendocrine Immunology and Behaviour the chief of the section, Esther Sternberg, summarized the different levels of analysis being carried out. These ranged from the systems level (e.g. neuroendocrine responses) to the neuroanatomical level (e.g. peptide expression in the brain) through the cellular level (e.g. hypothalamic cell neurohormone and neuropeptide production) to 'the molecular level (glucocorticoid receptor, estrogen receptor, other nuclear hormone receptors, cytokine and cytokine receptors)'.[37] It is, of course, at this molecular level that the current mechanistic approach to biology brings the greatest possibilities for knowledge that can be used for good in medicine or ill in hostile applications.

We can see this molecular level of understanding developing, for example in regard to aging[38] or to autoimmune and inflammatory diseases.[39] As noted in previous chapters, what is of particular importance is the impact of the genomics revolution on our understanding of cellular receptors and receptor mechanisms in response to intercellular signalling molecules. There is obviously considerable medical interest in glucocorticoids because of their wide range of physiological functions both in relation to the stress response and a range of other vital endocrine functions. Researchers therefore face the problem of finding selective drugs which will activate one function only without causing side-effects due to other functions. Additionally, as well as glucocorticoid receptors there are other steroid hormone receptors which can react to the same glucodorticoids and therefore cause further effects. Consequently, 'the identification of more selective functional ligands remains a goal of clinical and pharmaceutical research'.[40]

The glucocorticoids act rather directly to regulate gene expression. The glucocorticoid receptor is one of a group of intracellular receptors. When the receptor is activated it binds with endogenous ligands to form a complex in the cytosol which is then transported to the nucleus. There it can act directly as a transcription factor or indirectly by affecting the function of existing transcription factors. It is believed that the anti-inflammatory effects occur via the indirect route. Medicinal chemists are obviously taking advantage of the fact that even subtle changes in the structure of ligands can produce dramatic changes in receptor selectivity and, on the basis of this understanding, trying to find new drugs. They also are pursuing the goal of finding drugs which act selectively on particular tissue types. Importantly, whilst glucocorticoid research has been ongoing for half a century, only in the last decade has a basic understanding of the way in which the glucocorticoid receptor operates to regulate gene expression been gained. Considerable new advances therefore seem highly likely in the coming years.

The understanding of corticotropin-releasing factor (CRF) and CRF receptors has also undergone remarkable development in recent years.[41] CRF receptors, of course, belong to the class of G protein-coupled receptors. However, we now know that there are two different CRF receptors, designated CRFR1 and CRFR2. There are also related ligands, urocortin I (Ucn I) and urocortins II and III. CRF was first shown to be important in the stress response but these receptors and ligands are now known to be involved in a range of functions and to have a wide central nervous system and peripheral organ distribution.

It has been found that CRF itself has a ten-fold higher affinity for CRFR1 over that for CRFR2. However, Ucn I has equal affinities for both while Ucn II and III seem to be selective for CRFR2. The tissue distribution of the two receptor types in the brain appears to be rather clearly related to their physiological functions, with CRFR1 being located in anterior pituitary corticotropes and being stimulated by CRF to activate the release of ACTH. Direct infusion of CRF or Ucn I into the brain, or peripherally, produces a major release of ACTH from the pituitary and both peptide and small molecule antagonists of CRFR1 have been shown to blunt this response to stress in animals and humans. Many different types of knock-out mice have been produced to further confirm and investigate the mechanisms involved in the stress response.

To take one example, transgenic mice overexpressing CRF were produced to study the consequences of chronic HPA axis activation. As expected, these mice showed Cushing's syndrome-like symptoms (characterized by upper body obesity, weakened bones, severe fatigue, weak muscles, high blood pressure and sugar levels) because of the heightened ACTH and corticosterone production. On the other hand, mice deficient in CRF production showed that both basal and acute stress response levels of corticosterone were blunted in the absence of CRF. Similarly, mice lacking CRFR1 receptors showed a blunted response to stress (as measured through plasma ACTH and corticosterone levels). Thus there is a very clear understanding of the role of CRF/CRFR1 in the stress response. Evidence is also accumulating to suggest that CRFR2 activated by Ucn I, II and III may be involved in damping down the actions of CRFR1.

There is another burgeoning line of research activity which strongly suggests involvement of CRF and associated ligands and receptors in anxiety behaviours and depression:

> Strong evidence links stress and the sensitivity of the individual to stressful encounters to the development of depression ... a large body of evidence now ties CRF to the development of depression.[42]

Again, it is not at all difficult to see how such knowledge may be used to help people with depression – or for hostile purposes.

What is important to note is that our increasing understanding of molecular mechanisms (derived from the genomics revolution) is allowing a very rapid accumulation of new knowledge. For example, at the beginning of this chapter we discussed the bacterial toxin SEB, which has previously been weaponized. A closely related toxin, staphylococcal enterotoxin A (SEA), is also a super-antigen. A study published in 2002 showed that, like SEB, mice challenged by injection with SEA showed a dose-dependent increase in corticosterone and ACTH levels and behavioural effects such as an augmented rejection of novel food, indicating increased anxiety.[43]

Then, in a publication in 2005, it was shown – by the use of specific drugs known to block the different receptors – that these effects were most likely the result of CRF acting specifically on the CRFR1 receptor and not on CRFR2.[44] Most recently, it has been shown that this effect is dependent on the release of the cytokine tumour necrosis factor alpha (TNFα) following SEA administration. Normal mice injected with SEA were shown to have increased activity in relevant areas of the brain whereas mice unable to produce TNF (TNF$^{-/-}$ knockout mice) did not show this after challenge with SEA but still showed brain responses to other stressors. As might be expected:

> the augmented reduction in novel food intake after SEA challenge was attenuated in TNF$^{-/-}$ mice as well as in wild-type mice administered antibody to TNFα.[45]

Clearly, what is being elucidated is a circuit in which a toxin activates an immune response which then through changes in the nervous and endocrine systems produces, in part, a behavioural change. Such then is the power of our increasing understanding of agents that are likely biological warfare candidates!

3. Immune regulation of the nervous system

Infection is one of the main natural stimuli for modulation of the nervous system by the immune system. In this way, the immune system signals to the brain that there is a present danger of infectious microorganisms such as viruses, bacteria or animal parasites. Signalling is accomplished in one primary manner through the action of proinflammatory cytokines (cytokines causing inflammation reactions) such as interleukin 1 beta (IL-1β), tumour necrosis factor alpha (TNFα) and IL-6, which are produced by cells of the immune system after contact with microorganisms or their products.[46] The cytokines gain entry into the circulation from sites of the immune response in tissues and organs, and subsequently trigger reactions in the brain collectively known as sickness behaviour, which is characterized by fever, drowsiness, lethargy and loss of appetite.[47] The substances that are able to trigger the production of inflammatory cytokines are known as pyrogens for their ability to cause fever. Most notable among the pyrogens is lipopolysaccharide (LPS), known as endotoxin, which is a surface component of a large group of bacteria. The process of cytokine production by cells of the immune system has been discussed in Chapter 4, section 2.1 (Mammalian immune system).

The mechanism of fever induction has been studied most extensively. The brain is protected from the potentially harmful effects of biologically active substances or cells in the circulation by the blood-brain barrier, which is manifested by the extremely tight junctions between the endothelial cells lining blood capillaries,[48] which prevent circulating substances of a particular size and chemical property (as well as cells) from entering the brain. Proinflammatory cytokines are of a sufficiently large size that makes it unlikely that they can pass the blood-brain barrier. However, numerous pathways to explain how cytokines produced in the periphery can influence central nervous system events have been studied. For example, highly localized windows in this barrier known as circumventricular organs, which are sites that have blood capillaries with open junctions, allow passage of cytokines from the circulation into the brain. Evidence suggests that such windows are located in the anterior area of the hypothalamus.[49]

The cytokines entering the brain at this point bind to their receptors on cells in this area of the hypothalamus and induce them to produce the biologically active substance prostaglandin E_2 (PGE_2). It should be noted that many different cells of the central nervous system have receptors for virtually all known cytokines.[50] PGE_2 subsequently binds to its receptors on cells in the thermoregulatory center of the hypothalamus and induces reactions in neurons involving cyclic adenosine monophosphate (cAMP, a signalling molecule) and neurotransmitters to elevate the temperature set point.[51] Alternatively, it is also known that the proinflammatory cytokines interact with the endothelial cells lining the blood vessels of the hypothalamus and induce these cells to produce PGE_2, which can apparently cross the blood-brain barrier into the anterior hypothalamus. There is also indication that cytokines are actively transported through specific carriers across the blood-brain barrier. A further possibility in overcoming this barrier lies in the findings that afferent nerve fibres of the vagus nerve may transport inflammatory cytokines to the thermoregulatory centers of the hypothalamus.[52] In addition, there is evidence that the blood-brain barrier becomes more penetrable with an increasing immune response, and that *activated* T cells of the immune system can readily enter the brain parenchyma, whereas non-activated T cells are excluded under normal conditions.[53]

Virtually all known cytokines and their receptors have been found in many different types of cells in the central nervous system.[54] For example, IL-1β is produced locally in the brain by neurons, microglia and astrocytes.[55] Normally, there is little active production of proinflammatory cytokines in the brain itself due mainly to the fact that intact, functional neurons strongly suppress immune reactivity both in surrounding glia cells and in the neurons themselves.[56] Results of experiments in which neuronal bodies have been damaged or normal physiological input into innervated target tissue has been disrupted show that this coincides with the *de novo* expression of major histocompatibility complex (MHC) molecules of class I and II, which are conditionally involved in adaptive immune responses (see Chapter 4).

Further studies have shown local gene transcription leading to the production of IL-1β, TNFα and interferon gamma (IFNγ) occurred after neurons in the local vicinity were damaged. Thus, the brain exerts a tight control over local immune responses, but these can occur rapidly after regulating neurons are damaged, setting off a strong inflammatory response of microglial cells serving as cytokine producers of innate immune responses and also as antigen-producing cells for T lymphocytes of adaptive immune responses.

In addition to affecting the central nervous system through the action of cytokines, the immune system can also exert its effects on the nervous system through what is known as 'hardwiring', which consists of neural connections with lymphoid tissue in the autonomic nervous system. The autonomic nervous system connects the central nervous system directly to visceral target tissues, including those of the immune system, via sympathetic and parasympathetic nerves.[57] The parasympathetic nervous system pathways innervate lymphoid tissues via the neurotransmitter acetylcholine, and the sympathetic nervous system pathways innervate lymphoid tissues via the neurotransmitter norepinephrine. There are apparently extensive neural connections with the thymus, bone marrow, lymph nodes, spleen and gut-associated lymphoid tissues. Lymphocytes also have receptors for neurotransmitters in addition to acetylcholine and norepinephrine, including vasoactive intestinal peptide, pituitary adenylyl cyclase-activating polypeptide, calcitonin gene-related peptide (CGRP), substance P, histamine and serotonin. In addition, receptors for neuroendocrine mediators, including corticotropin-releasing factor (CRF), alpha melanocyte stimulating hormone (α-MSH) and leptin are found on lymphoid tissue.[58]

There are extensive intimate relationships between neural pathways that mediate pain and the immune regulation of inflammation. Afferent sensory nerves connect the peripheral organs such as the skin to the spinal cord. These nerves can be stimulated by tryptase released from mast cells, which are cells involved in allergic immune responses, resulting in the release of both CGRP and substance P from C-type sensory nerves in peripheral tissues and the spinal cord, which can both mediate pain. Neural release of CGRP and substance P also causes oedema and inflammmation in the skin.[59] Histamine is another neurotransmitter released from mast cells during allergic immune reactions and has been associated with the disease experimental autoimmune encephalomyelitis (EAE). Blockade of histamine 1 receptors as well as modulation of mast cell activity or deletion of mast cells can block the manifestation of EAE.[60] Furthermore, histamine 1 receptors are found in large numbers in brain lesions of multiple sclerosis.[61]

3.1. Malign manipulation?

In the preceding section possible ways in which the immune system can regulate functions of the nervous system were examined. Some of the consequences of using these regulatory elements with malign intent will now be discussed in more detail.

Proinflammatory cytokines can have several different effects on the nervous system. As may be recalled, these cytokines include IL-1β, TNFα and IL-6, all of which can apparently induce the state known as sickness behaviour, including fever, sleepiness, lethargy, loss of appetite and body weight loss. However, although IL-1β and TNFα do without a doubt contribute to the febrile response, the production of IL-6 is apparently crucial. This was seen in experiments with transgenic mice that were manipulated to overproduce a naturally occurring antagonist of the receptor for IL-1β (IL-1ra), which blocks IL-1β effects. Systemic injection of LPS still triggered a febrile response in these mice. Similarly, LPS could induce fever in mice lacking receptors for TNFα. However, neither LPS, IL-1β nor TNFα could induce fever in IL-6 deficient mice.[62] As for the other characteristics of sickness behaviour it is known that IL-6 can induce sleep, while TNFα is particularly effective in inducing loss of appetite and weight.

Thus, IL-6 is a particularly important cytokine considering its essential role in the induction of fever and lethargy, which represent the most debilitating characteristics of sickness behaviour. Nevertheless, TNFα and IL-1β can be considered just as significant in mediating sickness behaviour, because TNFα can induce the production of both IL-1β and IL-6, while IL-1β can induce both TNFα and IL-6.[63] In its medical research work on endogenous bioregulators the US Army has reported that IL-1 was effective in aerosol form in basic pulmonary absorption studies.[64] It therefore stands to reason that administration of IL-1β in aerosol form might indeed be an effective delivery system for inducing sickness behaviour in a population.

Another effect of IL-1β and the other proinflammatory cytokines on the central nervous system is the induction of the production of corticotropin-releasing factor (CRF) from the hypothalamus.[65] IL-1β is a major upregulator of CRF. In an earlier part of this chapter, it was discussed how CRF can lead to suppression of immune responses through its action on the pituitary to secrete ACTH, which in turns acts on the adrenal gland to induce the production of glucocorticoids. However, CRF has a profound effect on the nervous system as well. In this regard, overproduction of the hormone has been implicated with neurotoxicity and neurodegeneration in animal studies. For example, in an animal model of acute ischemia (stroke), it was shown that CRF blockers could protect against the loss of neurons which occurs as a result of a stroke. In addition, CRF has been implicated with major depression, anorexia nervosa and Alzheimer's disease.[66]

In Chapter 4 the characteristics and elements of the adaptive immune system were compared with those of the innate immune system. It was discussed how innate responses differ from adaptive responses because of the different components directing them. The immune system is involved in many illnesses associated with the nervous system, and one can see a clear division of these types of disorders, which are a reflection of the elements of immunity that are thought to be involved. Innate immune responses are associated with neurodegererative diseases such as Alzheimer's and Parkinson's diseases, in which proinflammatory cytokines and complement components are present in the central nervous system. On the other hand, specific antibody and T lymphocyte responses (adaptive immune responses) to acetylcholine receptors are seen in myasthenia gravis, while antibody responses to the glutamate receptor, which cause the blockade of glutamate-mediated synaptic transmission, are apparent in Rasmussen encephalitis, resulting in epilepsy. The occurrence of these diseases with the involvement of various components of innate and adaptive immunity shows that such elements can indeed have devastating effects on the nervous system, when the delicate balance betweeen activation and inhibition is broken.

When considering the possibility of using components of the immune system with malign intent, the question arrises as to the feasibility of targeting these elements to the central or autonomic nervous system. We have shown several examples in which the systemic application of bioregulators such as the proinflammatory cytokines is successful in overcoming the blood-brain barrier. As for an effective delivery system, we just have to recall the experiments of the US Army showing that the proinflammatory cytokine IL-1 in aerosol form could successfully be taken up through pulmonary absorption.[67]

Another possibility to be considered is the delivery of a substance through a vector, such as a bacterium. Apparently the former Soviet Union in its offensive biological weapons program experimented with the production of weapons designed to affect the central nervous system. One of the proteins studied was myelin basic protein (MBP), which is a substance used in the investigation of experimental autoimmune encephalomyelitis (EAE). EAE is mediated solely by T lymphocytes and can be induced by immunization with MBP. In the animal model used to study this disease, animals develop cellular infiltration of the myelin sheaths surrounding and insulating neuron axons of the central nervous system within 2–3 weeks after immunization. The T cells cause demyelinization (dissolution of the protective sheath), which leads to paralysis.[68] Apparently Soviet scientists were successful in transferring the gene encoding the MBP into *Yersinia pestis*, the causative agent of plague.[69] Presumably, upon infection with the bacterium, the gene would be activated, MBP would be produced and initiate an autoimmune immune response. This agent was thus conceived as a super-weapon designed to cause two devastating diseases if it would work. What experiments were actually carried out was not reported.

A further possibility is to deliver the substance through a viral vector. It is interesting in this respect that the substance leptin, which is a protein involved in a negative feedback loop to control body weight, was delivered in animal experiments to the hypothalamus by placing the gene encoding the protein into adeno-associated viral vectors.[70] After injection of the virus into the hypothalamus, the gene was activated and the protein produced in infected cells. This method of delivery was thought to be necessary, because leptin is not efficient in crossing the blood-brain barrier. The use of leptin is relevant for our discussion, as leptin apparently increases the levels of IL-1β in the hypothalamus.[71] The direct injection into the hypothalamus is not a feasible delivery method for a biological weapon, but the possibility of manipulating a viral vector to gain neural tropism might not be too far-fetched. It should also be recalled that an ongoing immune response[72] and even stress itself can increase the permeability of the blood-brain barrier.[73]

4. Summary

We talk of the nervous system, the endocrine system and the immune system as if they were separate entities. This mode of analysis comes about because of the different pathways by which neuroscience, endocrinology and immunology have developed within the history of biology. In fact, of course, in the living organism these systems are thoroughly integrated in order that the animal (or human) functions in a unified way. Indeed, it is rather surprising that in recent years there has been so much surprise at the findings that demonstrate how closely the immune system is linked to the other two systems. There is a fine network of checks and balances exerted on the operation of all three systems by the elements within each of them. The perturbation of the function of one system will invariably have profound effects on the operation of the others. All three systems are interconnected through the hypothalamus-pituitary-adrenal axis via cytokines, hormones, neurotransmitters, peptides and their receptors, and also through innervation of neural and lymphoid organs and even cells of the immune system themselves.

Here then is an ideal target for those with malign intent. Clearly, if the complex balancing feedback system briefly illustrated in this chapter could be disrupted by the use of a bioregulator, an ideal method of incapacitation would be available to an attacker. To give an example, a selective overproduction of proinflammatory cytokines by cells of the immune system could easily tip the balance over to the negative side, with detrimental effects on both the immune and the neuroendocrine systems. Some of the reactions that might result would include a severely debilitating sickness behaviour – high temperature, feelings of malaise, muscle aches and pains – as well as significant immune suppression that would make a person acutely susceptible to infection. In addition, the overproduction of CRF that would occur in reaction to the cytokines could cause incapacitating depression. Such reactions would critically weaken a person's resistance to an attacker with malign intent.

From the point of view of potential malign manipulation, it follows that there is necessarily a new level of complexity. If malign manipulation of one system can affect two or three systems the defender's problem of diagnosis and treatment increases out of all proportion to the attacker's effort.

7
Assessing the Adequacy of the CBW Prohibition Regimes for the Challenges of the 21st Century

1. Introduction

Multilateral prohibition regimes are more than the legal texts – the BWC and the CWC in our area of concern – on which they are based. In addition to this legal dimension, the concept of 'international regimes' captures the political dimension of states acting on their own (on the domestic level) and interacting with one another (on the international level) in the implementation of these legal arrangements. As states participate in international regimes out of their own free will there is an expectation that they will strive to comply with the stipulations of the regime. In addition, as these regimes are often created to overcome collective action problems, i.e. situations in which individual state action to address a problem would yield sub-optimal results, states participating in a regime can be expected to have an interest in adapting a regime when the character of the underlying problem, which led to the regime's creation in the first place, changes.

The standards according to which arms control regimes are negotiated have witnessed a considerable evolution over time. The differences between arms control regimes that were established in the late 1960s and early 1970s – like the nuclear non-proliferation regime and the BW prohibtion regime – and those established in the 1990s – like the CW prohibition regime – are quite striking. Certainly the bar for what qualifies as 'best practice' in setting up arms control regimes is much higher today than it was thirty or forty years ago. This best practice standard finds its expression in a number of features of the regime like the availability of verification measures, or, alternatively, procedures to provide transparency, the 'bindingness' of the agreements underlying the regime, the strength of the norms that are guiding state action, the provision of sanctioning mechanisms (in cases of non-compliance), the universality of the control measures agreed upon, and the adaptability of the regime structure to changing circumstances in the issue area the regime has been set up to regulate.

Although all of these features impact on the effectiveness and robustness of the CBW prohibtion regimes, it is the last of the above characteristics, i.e. the regimes' capacity to adapt to S&T change, which we are especially interested in. We therefore first summarize the current revolution in the life sciences, thus broadening again our discussion of the selected areas presented in the preceding three chapters. This will be contrasted with the evolution of the two prohibition regimes which takes place in slow motion.

2. The life sciences on the fast track: S&T challenges for upholding the prohibition regimes

2.1. The biotechnology revolution

2.1.1 The historical context

In order to assess the likely future of the ongoing revolution in biology and its potential misuse for hostile purposes it is useful to briefly examine the history of biological warfare. Although there are a number of possible uses of biological warfare in the historical record, it is, in fact, difficult to substantiate most of these from the available evidence. An exception is the intentional use of smallpox by the British against North American Indians in the middle of the eighteenth century. Here there is clear documentary evidence in the British records and enough knowledge of the transmission of the disease for the method used – the donation of infected blankets – to have been a viable means.[1]

However, it was not until the advent of the 'Golden Age' of bacteriology at the end of the nineteenth century that a scientific understanding of the nature of infectious diseases began to be acquired.[2] The large majority of 'cases' of bioterrorism in the twentieth century remained in the realm of wishful thinking hoaxes or failed attempts,[3] but there was a series of successful offensive biological weapons programmes in major states. These began as early as World War I with both sides attempting to use agents such as anthrax against the valuable draught animal stocks of the other. This was clearly only possible through the understanding of bacteriological agents gained from the work of major scientists like Pasteur in France and Koch in Germany.[4]

During the interwar years there were also some major offensive biological warfare programmes, notably the massive and gruesome Japanese programme which resulted in many attempts to use biological warfare in China. However, the important change that came about was a growing understanding of aerobiology, and it was the British, fearing biological attacks from Germany, who discovered that the best way to attack humans and animals was by spreading agents on the air so that they were inhaled and retained in the lungs. This remained the most feared method of attack throughout the rest of the century and was the basis for many of the concerns about biological agents being used as weapons of mass destruction. Following World War II – as the early Cold War developed – the huge American offensive biological warfare programme used the growing capabilities for microbial production to produce the large quantities of agent required. And it is now known, of course, that in the later Cold War period the USSR was beginning to apply the emerging genetic engineering technology in its even larger offensive biological weapons programme.

From this history it is clearly not difficult to see that in the past the latest advances in biotechnology have been used in the offensive programmes of the day. Equally clearly, it can be expected that precisely the same thing is likely to happen in the future unless major efforts are made to prevent that occurring. The possibility of inadvertently opening up new avenues for misuse of the life sciences by both state and sub-state actors has recently been added to this classical misuse scenario.

2.1.2. Experiments of concern

Rather than being centred on the deliberate misuse of biology, for example in state-level biological weapons programmes, these new concerns are centered on scientists inadvertently doing experiments and publishing results that might be of assistance to those with malign intent. As we have discussed in Chapter 4 the first major furore over such 'experiments of concern' related to the efforts of Australian scientists to find a more effective way of dealing with the plagues of mice that can be very damaging there. The scientists sought to genetically engineer the gene for a mouse egg protein into the genome of the mousepox virus.[5] They expected that this would lead to an antibody response in infected mice against their own eggs and thus to rejection of the eggs. This contraceptive worked, but not well, so the scientists tried to enhance the immune response by also adding the gene for the cytokine IL-4 to the mousepox. Unfortunately, they discovered that this manipulation produced a highly lethal pox virus because the cell-killing arm of the immune response in the mice was shut down and the virus was able to replicate without this hindrance. It was obvious that other pox viruses could be manipulated in the same way, perhaps with similar results.

Other experiments which gave rise to concern included work on the synthesis of viruses[6] and on the mechanisms by which smallpox interrupts the operation of its host's immune system.[7] However, when some thought was given to the discussions provoked by these individual experiments, and their publication, a deeper problem quickly emerged. For example, in regard to the mousepox experiment, it seems that the same experiment was carried out on rabbitpox and then, by other experimenters, on cowpox (which can jump species). Worse still, there were rumours that it was being contemplated that the experiment should be checked out in monkeypox.

The net result was that questions began to be asked – not about individual experiments – but where the whole area of research was likely to lead in the years ahead.[8] At the same time, eminent scientists such as George Poste were insisting that it was thinking far too narrowly to regard the threat as only coming from infectious disease manipulation.[9] Poste urged thinking 'beyond bugs' to include the possible manipulation of the human body's many regulatory systems, arguing that as we come to understand the chemical circuits involved we will simultaneously gain the ability to manipulate them – for good or evil purposes. Furthermore, Matthew Meselson pointed out that there is no end in sight for this process.[10] Advances in biology are likely to continue throughout the decades of the coming century and there will inevitably be a constant threat of their misuse for hostile purposes.

Yet despite these clear warnings that there is a need to think beyond single experiments, microbiology and immediate events, few seem to grasp the full impact of ongoing advances on the potential threat – particularly as those advances are being driven by clear industrial interest in exploiting the new biology for peaceful purposes around the world.[11] In order to grasp what is really happening to biology it is therefore necessary to step back from the world of security concerns about bioterrorism and try to assess what is happening at the heart of biology following the successful completion of the Human Genome Project. Only then can we gauge how far and how fast the advances are likely to go in coming decades.

2.2. Towards biology as a mechanistic science and biotechnology as real engineering

The fundamental mechanism of evolution was elucidated by Charles Darwin some one hundred and fifty years ago. It took another one hundred years for the structure of the hereditary material – DNA – to be determined by Watson and Crick and a further fifty for the structure of the human genome to be described. Now, however, discoveries at the core of biology are coming thick and fast, and this is likely to accelerate discoveries in other areas of the discipline. So to understand how the biotechnology revolution might evolve we have to consider what has followed on from the Human Genome Project (HGP), understanding, of course, that the HGP moved biology into a new era of 'big science' with huge sums of money being spent by international consortia to achieve the designated goal.

2.2.1. DNA synthesis

Craig Venter, one of the leaders of the project to sequence the DNA of the human genome, is reputed to have summarized the next stage of the progress of biology by joking that '[h]aving cracked the genome, evidently it's time to start stitching it up again'. What he was referring to was the need biologists now see to greatly enhance their capabilities for synthesizing long sequences of DNA. Though they have had the capabilities to synthesize short stretches for decades, and though Wimmer's group synthesized the 7,500 nucleotides of the polio virus in 2002, the perceived need now is to go well beyond those limits. The problem is that it is costly and time-consuming to produce long strands of DNA and there is an unacceptably high error rate. However, new methods for synthesis, for example on microchips,[12] and new methods of error correction, for example by protein-mediated error correction,[13] are thought likely to reduce both the cost and time involved and the error rate. As we shall see, these capabilities will allow biologists to investigate effectively many critical processes, but they also have malign implications, for example in increasing the ability to synthesize viral pathogens. Some of the people involved in this work have recognized the dangers and suggested the need for better controls.[14]

Other techniques related to the manipulation of DNA are also advancing rapidly. For example, DNA shuffling – the breaking up of DNA sequences and recombination of the fragments into many related versions – allows for the artificial acceleration of 'evolution'. Thus it was recently possible, for those seeking a model system to further benign research, to quickly evolve an HIV strain that would infect the previously resistant cells of macaque monkeys.[15] A virus was thereby constructed to attack a new host. Similarly, the very recently discovered technique of RNA interference (RNAi) is being understood as a fundamental feature of cells that could lead to numerous clinical applications requiring the specific silencing of genes – for example, to stop the growth of a tumour.[16] Once again, malign applications are far from unimaginable.

Fortunately, more biologists are becoming aware of the dangers, one virologist warning that '[t]here is every reason to believe that the ... recombinant DNA techniques can be used to render this H5N1 [influenza] virus transmissible from humans to humans'[17] and thus create a deadly new epidemic.

2.2.2. Synthetic biology

Biologists wish to complement their DNA sequencing capabilities with better DNA synthesis capabilities, particularly so that they can test out hypotheses at the micro level about fundamental processes such as how biochemical networks function.[18] The aim is to be able to develop artificial models of how circuits work in nature and then to test out the predictions from the models in circuits that have been built in living tissues.[19] Clearly, this enterprise involves mathematicians and engineers as well as biologists, but surprisingly rapid progress can be made in seemingly complex problems such as the computational design of artificial ligands.[20]

2.2.3. Systems biology

If synthetic biology is working at the micro level to understand how the parts of living cells are constructed and operate together, systems biology comes in at the other end and takes a macro level approach. The scientists involved ask how all of the parts of the body operate as a whole. Again, this is work that requires the linking of the expertise of mathematicians and engineers with that of biologists. Systems biology has been described as:

> an emerging field that is characterised by the application of quantitative theoretical methods and the tendency to take a global view of problems in biology.[21]

Clearly, this is not an entirely novel view in biology, but widespread acceptance of the value of the approach for studying dynamic behaviour in complex networks indicates a major change in the field.

It is not just that systems biology offers the promise of a genuinely mechanistic biology which is at the root of this change. There are very sound applied technological reasons for the approach to be supported by industry. Drug development is very costly and involves many decisions about which drugs to develop. To the extent that biochemical pathways can be understood as systems, the rational basis for such decisions is strengthened.[22] So we should expect drug companies to encourage the growth of systems biology.

2.2.4. Beyond bugs

By standing back a little from our direct concerns with bioterrorism and biowarfare we can also see the value of George Poste's famous exhortation[23] to his colleagues to think well 'beyond bugs'. It is not just that what is happening at the core of biology affects all other areas of the discipline, but that what happens in these other areas also impacts on the core. Thus the genomics revolution allowed the structure of many proteinaceous cellular receptors (e.g. for neurotransmitters) to be elucidated, but now that knowledge enables us to more rapidly investigate how neuronal circuits function (systems biology). The overall effect is reflected in the UK's 2001 contribution to the BWC background paper. There is presently a real feeling of accelerating discoveries across many areas of biology as well as at the core.

One way to think about the future of biology is to consider an analysis that takes in the global network of biochemical interactions in cells and then is able to zoom down to *specific* biochemical pathways in that global network and then down to the interfaces of the interacting proteins within a particular *part* of a biochemical pathway.[24] We can gain an idea of the overall rate of change in biological discoveries by considering what we now know of the protein-protein interactions. Ten years ago the number of different protein types was estimated from genomic data to be about 1,000, about a tenth of which were known. Progress in structural biology over the last decade has led to knowledge now of almost the complete set of 1,000.[25] It has been estimated from a variety of data sets that there could be up to 10,000 types of protein interactions in nature. Today we know about 1,800 of these and the number is growing at a rate of 200–300 per year, but it is confidently expected that technical advances and new initiatives will greatly increase the rate of discovery. Thus soon:

> A structural repertoire of interaction types, combined with accurate protein interaction networks and electron tomography will ultimately provide atomic details of complex cellular processes.

Biology is indeed changing rapidly into a mechanistic science.

3. Prohibition regimes evolving in slow motion: undermining regime adequacy?

3.1. The BW prohibition regime

As noted at the beginning of this chapter, there is more to international regimes than the treaties they are often based on. In case of the BW prohibition regime which does not have at its disposal an international organization to oversee implementation of the regime's standards for behaviour, the five-yearly Review Conferences assume critical importance in this regard. These Conferences offer an important opportunity to take stock of the BWC's implementation and its final declarations provide useful clues for the shared interpretations of regime members on the status of regime adequacy.

In Chapter 3 we made the point that successive review conferences have found the misuse of S&T advances in the life sciences to be covered by the scope of the BWC.[26] What has changed dramatically over time, however, is the assessment of the misuse potential of new scientific developments. For the First Review Conference the three

Depositary States – United Kingdom, United States of America and the then Soviet Union – produced a joint paper on relevant scientific and technical developments.[27] This was divided into seven sections, covering *inter alia* recombinant DNA techniques, new infectious diseases and microbial control of pests. The assessments made in the paper appear today to have been somewhat optimistic. In regard to the new recombinant DNA techniques, for example, the paper states that:

> now and for the foreseeable future, development and production of fundamentally new agents or toxins would present a problem of insurmountable complexity ...

and more generally that:

> Although recombinant DNA techniques could facilitate genetic manipulation of micro-organisms for biological or toxin warfare purposes, the resulting agents are unlikely to have advantages over known agents sufficient to provide compelling new motives for illegal production or military use in the foreseeable future.

How many of us would agree with such sentiments today?

In the section on infectious diseases, which considered Marburg, Ebola and Lassa fever, the conclusion was reached that there are no 'current technical reasons for regarding these diseases as posing a new biological warfare threat' and as far as biocontrol was concerned, the paper again had an optimistic perspective:

> misuse of both expertise and facilities is adequately covered by the terms of the Convention and this risk appears to be outweighed by the significant peaceful potential in this method of pest control.

The paper's general conclusion was in line with these specific ones:

> From a scientific and technological standpoint, the developments discussed in this paper, which are directed to peaceful purposes, do not appear to alter substantially capabilities or incentives for the development or production of biological or toxin weapons.

This, then, is our baseline assessment. The 1980 paper considered there was little to be greatly concerned about.

For the Third Review Conference in 1991 a number of states parties (Australia, Czechoslovakia, Sweden, the UK and the USA) produced contributions for the background paper on scientific and technological developments.[28] The UK contribution is particularly useful for our purposes since it used the same sections as the 1980 paper by the Depositary States. It therefore offers the possibility of directly comparing the conclusions.

For recombinant DNA techniques, now termed genetic modification (GM), the UK concluded that:

> in the period since the BWC entered into force the techniques of GM remain the most significant development among the scientific and technological activities that have relevance for the BWC ...

but, worryingly, it added:

> There has been steady refinement of those biotechnology aspects other than GM that an aggressor nation could misuse in developing an offensive BW capability; important among the capabilities that could be misused are techniques for the large-scale production of natural or modified micro-organisms or toxins ...

and it noted that further advances in such capabilities were to be expected.

In regard to new infectious diseases, again the perspective had changed:

> it must be recognised that the continuing increase in knowledge and expertise related to these newly recognised diseases and arboviruses in the public health context with the passage of time, can only increase the potential for misuse of such micro-organisms.

And similarly in regard to pest control:

> there has been increased study of factors relevant to effective dissemination. Such knowledge could in principle be misused by an aggressor intending to attack crops ... Some aspects of the dissemination technology would also be relevant to the deliberate release of organisms or toxins harmful to humans or animals.

As should be expected from such specifics, the general conclusion also changed.

While stressing that the BWC still covered all these developments and that some also had the potential to assist the defence, there is no doubting the changed view:

> The 1986 paper felt there was by then an increased potential for the large-scale production of BW agents with enhanced military utility. The current UK view is that worldwide the increase in knowledge of many of the pathogenic species of micro-organisms, and the knowledge of toxins and other biological agents, and the continuing pace of developments in civil biotechnology areas, have further increased the possibilities for production and hostile use of biological agents, whether naturally occurring or not.

So the situation had progressively worsened over successive five-year periods.

The general point was also made in the Australian contribution which, while noting the benefits brought by advances in production, harvesting and preservation of microorganisms, plant and animals cells, noted that 'it has also the potential, if misused, to provide the expertise and experience needed for developing and producing BW agents'. Furthermore, the Australians stated that because of these advances and their commercial value, many nations now had biotechnology capabilities that could be misused.

In its summary and conclusions section, Sweden's contribution also emphasized the speed of change:

> There has been a rapid progress in many areas of molecular biology and biotechnology in the period 1986–1991. Using molecular biology, mechanisms of virulence and infection have been identified and the same techniques may also permit deliberate manipulations of these mechanisms. Thus there is a potential danger that new or genetically modified BW agents may be created.

This Swedish contribution also stressed the growth and spread of industrial biotechology capabilities.

The changing scope of developments was emphasized by the Canadians who produced a special monograph, *Novel Toxins and*

Bioregulators: the Emerging Scientific and Technological Issues Relating to Verification and the Biological and Toxin Weapons Convention.[29] This monograph was circulated to all states parties at the Review Conference. The issue of peptide bioregulators was also covered in some detail by the United States:

> Their range of activity covers the entire living system, from mental processes (e.g. endorphins) to many aspects of health such as control of mood, consciousness, temperature control, sleep, or emotions, exerting regulatory effects on the body. Even a small imbalance in the natural substances could have serious consequences, including fear, fatigue, depression or incapacitation. These substances would be extremely difficult to detect but could cause serious consequences or even death if improperly used.

In general, the United States agreed on the speed of change:

> The past ten years have witnessed impressive strides in the fields of molecular biology and biotechnology. As the two juxtaposed words 'molecular biology' imply, the distinction between biology and chemistry is becoming blurred.

As in other contributions, the benefits to security and in defence are noted, but significantly the United States added:

> The confidence derived from the belief that certain technical problems would make biological weapons unattractive for the foreseeable future has eroded.

There is no doubt that there had been a major change in the perceptions of the contributors to these background papers between 1980 and 1991. However, the developments had been subject to a proper review in both 1986 and 1991 and agreed final declarations had been produced. The situation regarding review and agreed final statements deteriorated significantly thereafter. In 1996 there was less attention to other issues since the focus was on the work of the Ad-Hoc Group attempting to negotiate a legally-binding instrument to increase confidence in compliance, and in 2001–2002 the disruption caused by the United States prevented a final declaration being made. Nevertheless, it is possible to examine the background paper produced by states parties and to compare the 2001 version with that of 1991.

The background paper for the 2001 Review Conference had contributions from Bulgaria, South Africa, Sweden and the United States[30] and from the United Kingdom.[31] The large UK contribution will be discussed after a review of the other parts of the paper.

The South African contribution began by noting that there were many developments relevant to the Convention, but signalled its intention to deal just with biocontrol agents and plant inoculants. This was reasonable because anti-plant biological warfare possibilities are frequently neglected. It is also of interest because of the sanguine conclusions reached in 1980 at the First Review Conference. After a thorough review of these issues, the South African contribution concluded that there were many aspects of concern. For example, in regard to plant inoculants:

> A growing industry and more sophisticated production facilities that have the potential to be diverted to BW-producing facilities, as in the case of vaccine production facilities. ... The development of liquid inoculants that will make their application by spraying and aerosolisation a possibility.

This conclusion clearly differed from that of two decades previously.

Sweden began its contribution to the background paper with the observation that:

> The development within the field of biotechnology continues to be fast and innovative especially in the field of medicine. Part of this development is of concern to the BWC.

It added:

> Our understanding of the molecular mechanisms of microbial infections has increased immensely over the last decade.

In general, Sweden concluded that:

> Since the last Review Conference in 1996 the research in the field of biotechnology and molecular biology has entered a new and

> more complex era. Huge amounts of knowledge concerning basic principles of life have found worldwide applications. ... While these developments have been and are mostly beneficial they can also be misused.

Here Sweden appeared to be going along with the widely-held view that, in some sense, completion of the Human Genome Project signified the translation of biology and associated sciences into a new and more powerful state.

The contribution from the United States to the background paper is replete with references to the rapid developments in science and technology relevant to the BWC. In the opinion of the United States, for example in regard to bioinformatics:

> The first and most striking change in the last 5 years has been the amount of genetic information available worldwide ...
>
> Second, is the rapid increase in information technology that enables discovery of new constructs and their interrelationships to others on readily available low-cost computer equipment ...

and in regard to microbial genetics:

> Since the publication of the *Haemophilus influenzae* genome in 1995, the sequences of close to 30 microbial genomes have been completed during the past 5 years, and the sequences of more than 100 genomes, including several traditionally considered to be agents capable of being developed as biological weapons, should be completed within the next 2 to 4 years.

Again in seeming agreement with Sweden, the text states that '[s]cience, particularly in the biological and genomic areas, has advanced at incredible speed during the last 5 years, in large measure due to the stimulus of the Human Genome Project'. This makes sense as the project did move biology in the direction of 'Big Science' with huge funding and coordinated direction towards a particular goal. The point is made in the summary of the US contribution where it is indicated that the advances in the biological sciences have been enabled by parallel advances in other sciences and 'large-scale, international collaborative efforts'.

The UK clearly put a great deal of time and effort into producing its 29-page contribution to the background paper. One statement stands out as a general perspective:

> Throughout the various studies and consultations carried out by the UK to inform this review, it has been clear that the rate of change in science and technology fields relevant to the BTWC has been much greater than in the previous five-year period, that is between the third and fourth Review Conferences.

The text continues:

> A number of advances in scientific knowledge and its applications could be of consequence for the provisions of the BTWC. *Given the accelerating pace in science and technology, the UK wonders whether it is prudent to maintain a five-year gap between such assessments under the BTWC.* (emphasis added)

As with the US contribution, it is not difficult to find words like 'explosion' and 'burgeoning' in regard to the growth of the developments discussed.

Significantly, the UK text just quoted continued with a specific practical proposal:

> The UK suggests that the upcoming Review Conference consider *establishing a mechanism for State Parties to work together on a more frequent basis to conduct such scientific and technical reviews* and to consider any implications at the necessary level of expertise. (emphasis added)

Unfortunately, it would appear that this idea of designing a more adequate collective means of assessing and responding to scientific and technological change was lost amongst so much else in the chaos of the 2001–2002 Review Conference.

These official documents were not widely known about at the time, but one US official document did receive widespread coverage. This Department of Defense document was titled *Proliferation: Threat and Response.*[32] In part, it gave much more detail on how genetic engineering might be misused to 'improve' biological agents for hostile purposes. It suggested, for example, that a benign microorganism might be altered so that it would produce a harmful toxin or bioregulator or that a pathogen might be altered so that it was resistant to antibiotics or difficult to detect with standard methods. Though some publications continued to air concerns during the late 1990s and at the turn of the century, these did not receive very wide circulation. However, the events of September 2001 in the United States significantly changed many people's appreciation of the dangers and new concerns began to be raised in public.

However, it is questionable whether these new concerns have been translated into an overall strengthening of the BW prohibition regime. On the multilateral level the Inter-Review Conference process has focused on a small number of selected issue areas in which a substantial amount of information has been produced by BWC states parties. However, to the extent that much of this information is related to domestic implementation of BWC provisions, these activities follow from BWC Article IV on national implementation. Yet, this stipulation of the BWC should have been followed by states parties three decades ago. So, at best the current exercise represents an attempt to catch up with BWC implementation as it is required under the Convention and as such could be providing a foundation on which to build future efforts to deal with the changing BW threat.

If we accept the notion that on average powerful states are likely to have a greater impact on regime development than smaller states – a notion which is strongly supported by the fact that the US administration terminated the work of the Ad-Hoc Group in 2001 and also provided the blueprint for the work programme that has guided BWC states parties from 2003 to 2005 – prospects are not too good that this foundation on which to build will be utilized to the fullest extent possible. Not only has the US been instrumental in bringing to an end the work of the AHG, but it has also increased dramatically biodefence activities in response to the elevation of bioterrorism in the spectrum of threat to US national security. In this context one recent study found that the effects:

> of [regarding] bioterrorism as the number one threat to US security ... have not been limited to the national security sector, but have led to the securitization of the public health sector and the biomedical research infrastructure in the United States as well.[33]

If one considers biodefence and BW arms control to represent two sides of the same coin in the fight against BW, this clearly indicates a fundamental shift in priorities away from the multilateral enterprise of regime building and strengthening towards an increased reliance on national preparedness. As a result, no strong impulses from the US for achieving and maintaining regime adequacy should be expected in the foreseeable future.

3.2. The CW prohibition regime

Taking an agnostic stance one might argue that the CW prohibition regime has been set up to primarily ensure that the existing CW arsenals are destroyed and that states parties do not contribute to the proliferation of such arsenals in non-possessor states. Since the CWC's entry into force in April 1997 CW stockpiles in the declared possessor states are being destroyed under international verification and no instances of proliferation among CWC states parties or with their help has been recorded. From such a narrow interpretation of the regime's purpose there is no need to worry about the regime's adequacy.

Proponents of a second view would point to the CWC's drafters awareness of the fact that the CW prohibition regime has to operate in a dynamic scientific and technological environment and that therefore the CWC contains tools to make its states parties aware of S&T change – in the form of the scientific advisory board – and to address it – either collaboratively between the OPCW's Technical Secretariat and member states or during a regular Review Conference. A number of CWC state parties would certainly subscribe to such a view and as outlined in Chapter 2 these structures and mechanisms have been put in place and are operating. However, the portfolio of the SAB seems to be a rather limited one and the only full-scale assessment of S&T advances of relevance to the CWC was undertaken prior to the first CWC Review Conference. If this exercise is repeated for the next CWC Review Conference, which is to be held no later than the spring of 2008, the CW prohibition regime will fall into the same five-yearly pattern of assessing S&T advances BWC states parties have displayed in the past. What is more, with few exceptions the work of the SAB seems to focus more on technical problems related to the efficient operation of the CWC than a proactive scanning of S&T developments that might pose a problem to the regime a few years from now. One exception to this pattern was already mentioned in Chapter 3 and because of its significance is worth repeating here. The SAB in its report that was submitted to the First Review Conference by the Director General pointed out that it:

> was aware of concerns about the development of new riot control agents (RCAs), and other so-called 'non-lethal' weapons utilising certain toxic chemicals (such as incapacitants, calmatives, vomiting agents, and the like). ... The SAB noted that the science related to such agents is rapidly evolving, and that results of current programmes to develop such 'non-lethal' agents should be monitored and assessed in terms of their relevance to the Convention. However, based on past experience and the fact that many of these compounds act on the central nervous system, it appears unlikely from a scientific point of view that compounds with a sufficient safety ratio would be found.[34]

With the various breakthroughs in neurology over the past decade that have been fuelled by the genomics revolution proponents of such new toxic incapacitants might argue that they will be better able to specifically direct their new compounds and avoid 'collateral damage' among the targets being exposed to the so-called non-lethal CW. However, as our discussion of the misuse potential of S&T advances in the neurosciences in Chapter 5 has shown, the increase in knowledge about the human nervous system and its functioning as a system is very likely to lead to 'improved' chemical compounds which can be used for malign purposes to have a precise effect on human behaviour. Our own research thus confirms the SAB's assessment that the science underlying toxic incapacitants is evolving rapidly. Unfortunately, the SAB's call for monitoring and assessing these programmes up to now seems to have fallen on deaf ears on part of CWC states parties. Given the skirmishes during the First Review Conference over reorienting the industry inspection regime towards OCPFs which are producing PSF-related chemical compounds, the reluctance or rather disinterest in tackling more challenging and also more abstract S&T advances does not come as a surprise. However, should this attitude on part of influential CWC states parties prevail, it does not bode well for the prospects of the CW control regime to be kept up to date and thus be adequate to deal with the challenges ahead.

8
Conclusion: Towards an Overarching Framework for Biochemical Controls

1. Introduction

The evidence from the life science laboratories is quite clear: there is going to be an increasing risk that new discoveries will facilitate both state-level offensive biological weapons programmes and sub-state (terrorist) development of biological weapons. For over a decade it has been clear that only a wide-ranging integrated web of policies will be adequate to prevent this misuse of our new scientific and technological capabilities taking place. The web of deterrence[1] or web of prevention consists, at the very least, of:

> comprehensive, verifiable, global CB arms control to create a risk of detection and a climate of political unacceptability for CB weapons;
> broad export monitoring and controls to make it difficult and expensive for a proliferator to obtain necessary materials;
> effective CB defensive and protective measures to reduce the military utility of CB weapons; and
> a range of determined and effective national and international responses to CB acquisition and/or use.[2]

Thus, the proposed web of prevention encompasses the two CBW prohibition regimes, but goes beyond them in that it includes additional defensive, counterproliferation and supply-side measures like export controls and the interdiction of shipments that might contribute to CBW proliferation. What it does not account for, however, is the paradigm shift we outlined in Chapter 1 and which has guided the selection of the areas of S&T advances we have focused our attention on: the shift from CBW agents and possibilities of their manipulation towards the malign manipulation of selected physiological targets in the human body. In light of this paradigm shift it might no longer suffice to make all elements of the web as strong as possible, to persuade any proliferator contemplating the development of chemical or biological weapons that it is more likely that the potential costs far outweigh any benefits.

The question thus becomes how to achieve the appropriate strengthening of the various policy elements in the web *and* how to devise an overarching framework that would tie together all the measures that have been proposed and that will be needed additionally to account for the paradigm shift allowing future biochemical warfare. Given the slow and difficult progress of recent years it is safe to assume that no quick fixes will be available. Rather, a slow and iterative process of adapting and expanding existing prohibitions and controls both at the national and international level will be required.

2. National measures

Both the CWC and BWC have the requirement that necessary national measures are enacted to implement the treaties within States Parties, but the implementation of this requirement has been unsatisfactory for both agreements. With its organization and regular meetings the situation in regard to the CWC is beginning to be rectified through the operation of a specific action plan.[3] Investigation revealed that the reasons for failure to implement the national legislative requirement were frequently administrative or bureaucratic, not political, so appropriate assistance could greatly assist in effective rectification of the deficiency. The BWC has neither an organization nor regular meetings so development and implementation of such an action plan are far less likely. It may be that implementation of Security Council Resolution 1540 (to be discussed under international measures below) will have some effect on BWC national implementation, but this cannot be guaranteed.

Article IV of the BWC specifically states that:

> Each State Party to this Convention, shall, in accordance with its constitutional processes, take any necessary measures to prohibit

> and prevent the development, production, stockpiling, acquisition or retention of the agents, toxins, weapons, equipment and means of delivery specified in article I of the Convention, within the territory of such State, under its jurisdiction or under its control anywhere.

Thus it was perfectly appropriate for the first (2003) meetings of the Inter-Review Conference process to consider:

> i) the adoption of necessary national measures to implement the prohibitions set forth in the Convention including the enactment of penal legislation.[4]

And for the states parties to agree the value of the following:

> To review, and where necessary, enact or update national legal, including regulatory and penal, measures which ensure effective implementation of the prohibitions of the Convention.[5]

The problem, of course, is that the Inter-Review Conference process had a mandate only to 'discuss and *promote* common understanding and effective action' (emphasis added) on the topics under consideration.[6] Only at the Sixth Review Conference will the states parties 'consider the work of these meetings and decide on any further action'.

Given the increasing concern over misuse of the life sciences by sub-state actors since the initial formulation of the idea of a web of deterrence in the early 1990s, the second topic for discussion at the 2003 meetings was also understandable. This was consideration of:

> ii) national mechanisms to establish and maintain the security and oversight of pathogenic microorganisms and toxins.[7]

Again the states parties sensibly agreed on the value of:

> The need for comprehensive and concrete national measures to secure pathogen collections and the control of their use for peaceful purposes. There was a general recognition of the value of biosecurity measures and procedures, which will ensure that such

> dangerous materials are not accessible to persons who might or could misuse them for purposes contrary to the Convention.

No doubt, in a number of countries such as the UK, considerable efforts have been and are being made to achieve such goals.[8] But the fact remains that consideration of concerted international action will only be decided at the 2006 Sixth Review Conference of the BWC.

In considering disease monitoring it was recognized *inter alia* that:

> strengthening and broadening *national* and international surveillance, detection, diagnosis and combating of infectious disease may support the objective and purpose of the Convention;[9] (emphasis added)

and:

> the primary responsibility for surveillance, detection, diagnosis and combating of infectious diseases rests with *States Parties*. (emphasis added)

Regarding capabilities for responding to cases of alleged use of biological weapons, they also recognized that:

> States Parties' national preparedness and arrangements substantially contribute to international capabilities for responding to, investigating and mitigating the effects of cases of alleged use of biological or toxin weapons or suspicious outbreaks of disease.

However, in a step forward from the 2003 meeting of states parties, the 2004 report also stated that 'State Parties are encouraged to inform the Sixth Review Conference of, *inter alia*, any actions, measures, or other steps that they may have taken on the basis of the [2004] discussions.'[10] This will again clearly facilitate the process of strengthening of regime implementation at the national level.

2.1. Codes of conduct for life scientists

In 2005 the Inter-Review Conference process turned to the very different topic of:[11]

> v. the content, promulgation, and adoption of codes of conduct for scientists.

This clearly raised issues which were well outside the usual range of discussions at BWC meetings, brought in a new constituency of interest – life scientists, and brought much more clearly into focus the impact of the ongoing scientific and technological revolution.

Fortunately, the complexities of the subject of codes of conduct were clarified at the meeting and as one detailed report noted 'there appears to be recognition of the value of a matrix of codes'.[12] These codes would be composed of:

> an overarching set of moral and *ethical principles* which might have wide applicability, a *code of conduct* which could give guidance and, at the more detailed level, an extension to an existing national *code of practice* which might set out steps that need to be taken as a regular process when any new work is being considered.[13] (emphases added)

Unfortunately, it also became clear from many sources that there was a major prior problem in that life scientists had little awareness of the problem of the dual uses – benign or malign – to which their work might be applied.

One empirical study of the views of life scientists in UK universities concluded, for example, that:

> There was little evidence from our seminars that participants:
> a. regarded bioterrorism or bioweapons as a substantial threat;
> b. considered that developments in life sciences research contributed to biothreats;
> c. were aware of the current debates and concerns about dual-use research; or
> d. were familiar with the BTWC.[14]

And this notwithstanding the fact that they were used to tough regulation of their work (for example, in regard to animal experimentation and biosafety), had full access to major journals where the issues were prominent and had English as their working – often first – language. They can therefore be considered as very much a best case. Other parts of the worldwide life sciences community are unlikely to be much better informed. So to make any progress on the development of codes of conduct within the various states parties to the BWC a major awareness-raising programme will have to be undertaken.

2.2. Oversight of research

In one country at least this awareness-raising is beginning in earnest. As previously noted, the United States National Academies, in response to growing concerns about the misuse of modern biology, set up a committee under the chairmanship of Gerald Fink to examine the issues involved. The committee report, *Biotechnology Research in an Age of Terrorism: Confronting the Dual-Use Dilemma*,[15] in part identified seven classes of experiments that it felt required prior review. The classes were those which:

— would demonstrate how to render a vaccine ineffective;
— would confer resistance to therapeutically useful antibiotics or antiviral agents;
— would enhance the virulence of a pathogen or render a non-pathogen virulent;
— would increase transmissibility of a pathogen;
— would alter the host range of a pathogen;
— would enable the evasion of diagnostic/detection modalities; and
— would enable weaponisation of a biological agent or toxin.

Furthermore, the report noted that other classes of experiments might also require review in the future.

The report favoured a voluntary scheme of self-regulation, but suggested that a national board be set up to oversee and develop the scheme. The US government accepted this idea and moved to found a National Science Advisory Board on Biosecurity (NSABB). The NSABB held its inaugural public meeting in Washington DC in mid-2005 and detailed explanations of what had been done were given to the BWC meeting of experts in Geneva.

In Chapter 3 we have pointed out that in response to worries about scientific publications like that on the mousepox experiment, the editors of a number of major scientific journals had agreed on a system of checking submitted papers for possible inadvertent assistance being given to terrorists. However, this had resulted in very, very few papers even being questioned, let alone modified or refused publication. In light of the questionable value of self-regulatory efforts by scientific publishers, the NSABB model was regarded by many as an important potential model for other states' national oversight systems in the future. The difficulties with the proposed system, however, should not be underestimated.

What was being proposed appeared to be a tiered review system in which experiments in the designated classes would be subject to local review and then, if there were difficulties at that level, the national system would become involved. A group at the University of Maryland, which had given considerable attention to such a tiered review system, has expressed a number of concerns. They suggested that the system had to cover all institutions involved in biotechnology research (including civil industry and biodefence), that it had to be based on law, not guidelines, and that an effective system needed to be international, not just national.[16]

Biodefence is a particular concern.[17] Legitimate biodefence is permitted under the BWC and needs to be appropriately strengthened as part of the overall web of prevention against the hostile use of modern biology. But biodefence is necessarily going to be carried out in areas of research which are related to offensive possibilities. Thus increases in biodefence work have to be carried out carefully to avoid misperceptions arising in other countries about what is being done. Certainly, there has been a huge increase in biodefence funding in the United States, including that for the National Institutes of Health which grew 'by over 3,200%, from $53 million in fiscal year 2001 to a record $1.8 billion (requested) in fiscal year 2006'.[18] This was always likely to raise questions.

It is generally agreed that the community of practising chemists and their professional societies played an important positive role in the negotiation of the Chemical Weapons Convention. This positive role has continued, with the International Union of Pure and Applied Chemistry (IUPAC) making a major contribution on scientific developments for the 2003 First Review Conference of the CWC. More recently, IUPAC had joined with the OPCW in its efforts to develop new educational aids to inform the profession of the importance of the CWC.

The BWC obviously lacks an organization like the OPCW, but what is less well known is that the disparate international professional societies of biologists lack an overarching organization like IUPAC.

Getting an international assessment from the world's biologists of present and future developments and what might best be done is therefore very difficult. The UK Royal Society has argued that there is a need for the BWC to have a Scientific Advisory Panel to carry out such assessments.[19] This is a sensible idea but, in view of the difficulties in agreeing anything to strengthen the BWC,[20] it might be that the individual national academies will have to pursue this idea as a cooperative venture outside of the Convention until the prospects for the international regime improve.

3. Adapting CBW controls on the international level

As outlined in Chapters 2 and 3, attempts to strengthen the BW prohibition regime through the negotiation of a legally binding compliance protocol and to adapt the CW prohibition regime to changes in chemical industry and S&T advances have either been a complete failure – in the case of the AHG work on a BWC Compliance Protocol – or seen rather limited success – in the CW realm. The new Inter-Review Conference process established by the last BWC Review Conference in 2001/2002 is more focusing on national implementation measures in the selected areas that are being covered than genuinely moving the regime forward. For this reason the BWC Inter-Review Conference Process has been addressed already in the previous section.

However, there have been a few initiatives of an international character that have supplemented the two prohibition regimes in the areas of export controls and interdiction of NBC-related materials. Two of these will be discussed in the following section. This will be followed by a discussion of some proposals that have been made for strengthening the regimes through the negotiation of a biosecurity convention, the criminalization of CBW and the setting up of a small organizational infrastructure in support of the BWC. Yet, as all these proposals fall short of offering a coherent framework in which either their relationship to the CBW prohibition regimes is somewhat unclear, or they do not aim to integrate the various measures in existence and being proposed, or they do not take into account the paradigm shift with which we are concerned, we propose negotiation of a Framework Convention on Biochemical Controls (FCBC), a provisional outline of which will be presented in the final section of this chapter.

3.1. Measures already adopted

The Proliferation Security Initiative (PSI) was initiated by the US administration under President Bush in May 2003. It establishes a framework for 'multinational response to the growing challenge posed by the proliferation of weapons of mass destruction (WMD), their delivery systems, and related materials worldwide'.[21] It is based on the following principles to which all participating states subscribe:

- Undertake effective measures, either alone or in concert with other states, for interdicting the transfer or transport of WMD, their delivery systems, and related materials.
- Adopt streamlined procedures for rapid exchange of relevant information.
- Work to strengthen their relevant national legal authorities to accomplish these objectives and work to strengthen international law and frameworks.
- Not transport or assist in the transport of any cargoes of WMD, their delivery systems or related materials to or from countries or groups of proliferation concern.
- Board and search any suspect vessels flying their flags in their internal waters, territorial seas, or areas beyond the territorial seas of any other state.
- Consent under the appropriate circumstances to the boarding and searching of their own flag vessels by other states, and to the seizure of such WMD-related cargoes.
- Stop and/or search suspect vessels in their internal waters, territorial seas, or contiguous zones, and enforce conditions on suspect vessels entering or leaving their ports, internal waters, or territorial seas.
- Require suspect aircraft that are transiting their airspace to land for inspection and seize any such cargoes, and deny aircraft transit rights through their airspace.
- Prevent their ports, airfields, or other facilities from being used as transshipment points for WMD-related cargo.[22]

The Initiative started with 11 founding members and has so far attracted support from more than 60 additional states. In early 2004 the US signed two Ship Boarding Agreements with Liberia and Panama, respectively. This represents a significant step towards universalization of the PSI principles as roughly '50% of the world's shipping volume is carried on flag vessels of the core PSI participants, as well as Panama, and Liberia'.[23] Since then similar agreements have been signed with the Marshall Islands (August 2004), Croatia (June 2005), Cyprus (July 2005) and Belize (August 2005).

PSI participants have emphasized the compatibility of the initiative with the two CW and BW prohibition regimes. However, as one recent interpretation of PSI as a building block of an 'increasingly decentralized nonproliferation architecture'[24] shows, it represents only a partial solution to the wider biochemical threat we are facing: it focuses exclusively on one part of the supply side of the proliferation problem – as it does not address intangible technology transfers – and does not take into consideration the paradigm shift away from pathogens or toxic chemicals as weapons and towards the physiological targets in the human body.

UN Security Council Resolution 1540 of 28 April 2004[25] addresses some of the limitations inherent in the PSI approach, as first of all the export control measures required under it are not excluding intangible technology transfers. It also has a much wider scope than PSI prohibiting state support for non-state efforts to acquire NBC weapons, requiring states to:

> adopt and enforce appropriate effective laws which prohibit any Non-State actor to manufacture, acquire, possess, develop, transport, transfer or use nuclear, chemical or biological weapons and their means of delivery.[26]

Resolution 1540 furthermore requires all states to undertake 'appropriate effective measures to account for', 'effective physical protection measures' and 'appropriate effective border controls and law enforcement efforts' in order to secure NBC-weapons and related material and prevent their misuse by non-state actors. In addition it calls for 'appropriate effective national export and trans-shipment controls' and sets up a Committee of the Security Council to examine states' reports on the status of their implementation of Resolution 1540.

For states already participating in the CW and BW prohibition regimes this is largely a reiteration and specification of commitments already undertaken under the BWC and CWC. However, the wording in Resolution 1540 makes these stipulations binding on all states, including those outside the prohibition regimes and in that sense presents a useful step forward in ensuring universality of the norms against CBW. Yet, the wording of the resolution on the other hand also allows for loopholes in implementation. As it is not specified what constitutes 'effective appropriate' action, states might simply report back to the Security Council Committee set up under the resolution that they are already taking such action. Furthermore, while the Security Council expresses its intention 'to remain seized of the matter', the time limitation of the Committee's activities to two years presents an additional limitation on the positive effect the resolution might have. Especially when viewed from the perspective of S&T changes in the life sciences, a permanent institutional structure to examine states' control efforts would be much more appropriate to monitor and react to the changing threat spectrum.[27]

3.2. Additional measures proposed

A number of additional measures have been proposed by academics and NGOs to strengthen or supplement the CW and BW prohibition regimes. One proposal advocates the establishment of international biosecurity standards.[28] Like the work by the Maryland University group of scholars mentioned above, Tucker focuses on preventive measures that could be taken to better secure pathogens that could be misused by terrorists for a BW attack. He correctly points out the absence of:

> uniform global standards for laboratory security ... on which individual states can base national legislation and regulations. This lack of harmonization, ... has given rise to gaps and vulnerabilities that must be addressed as part of a coordinated global strategy to prevent bioterrorism.[29]

Elements of that strategy should include proper accounting mechanisms for pathogens and toxins, the registration and licensing of facilities handling pathogens and toxins, the establishment of physical security measures, and the screening of laboratory personnel.[30] Although UNSC Resolution 1540 seems to go a long way towards meeting these demands, the proposal put forward by Tucker contains much more detail in terms of the measures to be taken, aims at the harmonization of national measures, and provides for the establishment of an oversight mechanism which incorporates a small secretariat, both of which go well beyond the mandate of the 1540 Committee. Thus, agreement on more stringent international biosecurity standards along the lines proposed by Tucker could well serve as the next stage in the Security Council's dealing with the bioterrorist threat in general and the oversight of pathogenic microorganisms and toxins in particular. Clearly, the Security Council would not be the right venue to negotiate such a treaty on global biosecurity standards, but its initiation and support of a negotiation process leading to such a legally binding instrument could be expected to create a positive momentum for negotiations in the Sixth Committee of the UN General Assembly.

A different new international treaty which would make for a very useful addition to both BWC and CWC and thus improve regime adequacy is a 'Convention to Prohibit Biological and Chemical Weapons under International Criminal Law'. Such a convention has been proposed and a draft text formulated by the Harvard Sussex Program on CBW Armaments and Arms Limitation.[31] While both BWC and CWC require states parties to enact domestic legislation which criminalizes the prohibitions contained in these treaties, the new convention would put acts involving biological and chemical weapons on an equal footing with aircraft hijacking or torture. As pointed out by the authors of the draft convention, '[p]urely national statutes present daunting problems of harmonizing their various provisions regarding the definition of crimes, rights of the accused, dispute resolution, judicial assistance and other important matters'.[32] Not only do CWC and BWC lack such provisions in their requirements for states parties to implement the stipulations of the conventions, but the degree to which national implementation measures are enacted at all remains unsatisfactory to an extent that even in the CW realm CWC states parties at the First Review Conference in 2003 saw the need to initiate an action plan to increase compliance of states parties with the national implementation measures demanded by the Convention.[33] At its core under the proposed convention:

> [e]ach State Party would be required, *inter alia*, (i) to establish jurisdiction with respect to such crimes according to established principles of judicial law, including the principles of territoriality, nationality, protection, and passive personality, and (ii) where the state has jurisdiction and if satisfied that the facts so warrant, to

> submit those cases to competent authorities for the purposes of extradition or prosecution. Further with respect to actual use of biological and chemical weapons, each State Party would be required to establish jurisdiction over all persons found on its territory regardless of their nationality or place of the offence.[34]

While such new international treaties would enhance regime adequacy by increasing the norm and rule density in the issue area and thus complement BWC and CWC, other proposals are more closely linked to the BWC and have for example explored opportunities to address the 'institutional deficit' of the BWC.[35] In contrast to adding more legally binding obligations for regime participants, this would aim at improving the organizational capacities for better implementing already existing normative guideposts of the BW prohibition regime. Starting from the assumption that even after the collapse of the Ad-Hoc Group negotiations a comprehensive organizational structure like the proposed Organization for the Prohibition of Biological Weapons would represent the ideal solution for the BW regime's organizational deficit, Sims argues that:

> for the years immediately following the Sixth Review Conference some less ambitious proposals must suffice. These might comprise an annual meeting of States Parties, or an open-ended meeting of their Bureau, supported by a Scientific Advisory Panel and a permanent Secretariat. All would derive their authority from the Sixth Review Conference.[36]

This latter point is important as it makes clear that any such modest organizational setup would not require an amendment to the BWC, involving ratification by states parties. Sims builds a well-argued case in support of his proposal, quoting increasing support from BWC states parties during the past few years and also pointing to the nuclear non-proliferation treaty (NPT), where states parties have made similar proposals for establishing a mechanism for more regular reviews of the treaty than the five-yearly review conferences permit. However, although there is growing support for closing the institutional gaps in both the nuclear non-proliferation and the BW prohibition regimes, the fate of the 2005 NPT review conference does not bode well for the 2006 BWC review. As in the case of the NPT, it is not unrealistic to expect resistance on part of some states parties to increase formal interactions among BWC states parties between review conferences. Clearly the US government has made its unwillingness to engage in any such activities known. As decisions by review conferences are taken by consensus, the weakest link determines the strength of the chain, in this case the strength of any measures – institutional or otherwise – upon which the Sixth Review Conference will be able to agree.

3.3. Bridging the gaps and providing an overarching framework: towards a Framework Convention for Biochemical Controls (FCBC)

As this brief discussion of a few of the already enacted or at least proposed measures to strengthen the BW and CW prohibition regimes clearly demonstrates, all of these initiatives and proposals are addressing only a part of the problems identified and practically none of them addresses in a meaningful way the paradigm shift in the life sciences we have explored in previous chapters and have illustrated by discussing advances in the neurosciences and in immunology. As we have pointed out, however, adapting the CBW prohibition regimes to scientific and technological change is one of the key measures to ensure future regime adequacy. What is thus needed is an instrument to bridge the existing gaps and address the new challenges looming on the horizon of efforts to prevent the misuse of S&T advances in the life sciences.

We propose that such an instrument could take the form of a new Framework Convention on Biochemical Controls (FCBC). Framework conventions fall in the area of soft law which does not make legally-binding specific proscriptions for state action. They are a relatively new type of legal instrument which has been applied *inter alia* in the areas of international environmental and international health policy. The 1992 UN Framework Convention on Climate Change[37] and the 2003 WHO Framework Convention on Tobacco Control[38] are two examples of such conventions which aim at providing a wider framework for specific action to be agreed upon at a later stage. Given the lack of progress towards verification measures in the BWC context, it might be advisable to take a step back from the details of the two regimes and look at the larger picture of scientific and technological advances.

While a FCBC will not immediately address the structural shortcomings in the BW prohibition regime or improve compliance with the CWC's stipulations, its longer-term benefits would be considerable. They would be located in two areas: firstly, a FCBC could provide a bridge between issue areas that so far have been treated largely in isolation. This refers to both chemical and biological weapons developments, where an overlap in the subject matter, i.e. in regard to toxins, which is being regulated, is widely acknowledged, but where an overarching bracket is currently missing. While the separation of CW and BW made sense with respect to classical warfare agents and was useful in negotiating the BWC in the late 1960s and early 1970s, in the age of 21st century life sciences in which biological processes can be affected at the molecular level, the continued distinction will look increasingly out of touch with reality. The FCBC could also provide a stronger link between the prohibition of CBW and human rights. So far this link has been made only in relation to toxic incapacitants.[39] Secondly, a FCBC could address the paradigm shift from the chemical or biological warfare agent as the object of malign manipulation to the physiological target in the human body, thereby contributing to the continued adequacy of the CBW prohibition regimes.

In order to provide these benefits the FCBC could be used as a vehicle by states to recognize the dual-use potential biological and chemical agents and materials, equipment, technologies and know-how and to express their determination to prevent the misuse of chemical and biological agents and materials to incapacitate, kill or purposefully harm humans, animals or plants. An FCBC would provide an opportunity to acknowledge the accelerating speed of scientific and technological advances in the life sciences and related scientific disciplines and also to acknowledge the above mentioned paradigm shift in the proliferation problem as knowledge that could be misused to target the human body is widely diffused. From this vantage point then, the framework convention should reflect the concern that an increased understanding of life processes at the molecular level will amplify the misuse potential of biological and chemical agents and materials, equipment, technologies and know-how to kill, harm or otherwise incapacitate. The FCBC should make it clear that no person should be exposed to such biological and chemical materials without having given a prior informed consent to such exposure. States parties to the FCBC should be required to treat any violation of such a stipulation as a criminal act and make it punishable under national and international law. Last, but not least, the FCBC would have to be drawn up in a way that it does not contradict or detract from existing international treaties and other arrangements relating to chemical and biological agents and materials, most notably the CWC and BWC.

Notes and References

1 Introduction and Overview

1. See D. E. Kaplan and A. Marshall, *The Cult at the End of the World* (London: Hutchinson, 1996); M. Leitenberg, 'The Experience of the Japanese Aum Shinrikyo Group and Biological Agents', in B. Roberts (ed.), *Hype or Reality: the 'New Terrorism' and Mass Casualty Attacks* (Alexandria, VA: CBACI, 2000), pp.159–70.
2. See M. R. Dando, *Preventing Biological Warfare. The Failure of American Leadership* (Basingstoke: Palgrave Macmillan, 2002).
3. See Chapter 4 for a discussion of these experiments.
4. See M. A. Levy et al., 'The Study of International Regimes', *European Journal of International Relations*, 1 (3), 1995, 267–330; A. Hasenclever, P. Mayer and V. Rittberger, *Theories of International Regimes* (Cambridge: Cambridge University Press, 1997).
5. A. Hasenclever, P. Mayer and V. Rittberger, *Fair Burden-Sharing and the Robustness of International Regimes: the Case of Food Aid*, Tübinger Arbeitspapiere zur Internationalen Politik und Friedensforschung Nr.31, Tübingen, 1998, p.1, emphasis in original.
6. H. Müller et al., *Regime unter Stress. Beharrungs- und Anpassungsleistungen internationaler Regime unter den Bedingungen existenzgefährdender Herausforderungen* (Frankfurt/Main: PRIF, February 1999), unpublished manuscript.
7. Müller et al., *Regime unter Stress*, pp.10–11, 15–24.

2 Science, Technology and the CW Prohibition Regime

1. The major reason for the exclusion of chemicals that are toxic to plants from the CWC's definition lies in the United States' usage of defoilants during the Vietnam War and its refusal during the CWC negotiations to have this included in the definition.
2. World Health Organization (WHO), *Health Aspects of Chemical and Biological Weapons. Report of a WHO Group of Consultants* (Geneva: WHO, 1970), p.27.
3. On the problematic of so-called non-lethal CW agents see sections 2.5.3 and 3.1.
4. WHO, *Health Aspects of CBW*, p.28.
5. T. P. Noeller, 'Biological and chemical terrorism: Recognition and management', *Cleveland Clinic Journal of Medicine*, 68 (12), 2001, 1001–16.
6. See F. R. Sidell et al., 'Vesicants', in F. R. Sidell, E. T. Takafuji and D. R. Franz (eds), *Medical Aspects of Chemical and Biological Warfare* (Washington, DC: Office of the Surgeon General, 1997), pp.197–228.

7. See F. R. Sidell, 'Nerve Agents', in Sidell, Takafuji and Franz (eds), *Medical Aspects of Chemical and Biological Warfare*, pp.129–79.
8. See H. D. Crone, *Banning Chemical Weapons. The Scientific Background* (Cambridge: Cambridge University Press, 1992), pp.13–15; D. Martinetz, *Vom Giftpfeil zum Chemiewaffenverbot. Zur Geschichte der chemischen Kampfmittel* (Frankfurt/Main: Verlag Harri Deutsch, 1995), pp.7–54.
9. See Crone, *Banning Chemical Weapons*, p.17.
10. On the breadth of this industrial revolution and speed with which it developed in the second half of the 19th century see F. Aftalion, *A History of the International Chemical Industry. From the 'Early Days' to 2000*, 2nd edn (Philadelphia: Chemical Heritage Press, 2001), especially pp.32–101.
11. J. P. Robinson, 'The Negotiations on the Chemical Weapons Convention: a historical overview', in, M. Bothe, N. Ronzitti and A. Rosas (eds), *The New Chemical Weapons Convention: Implementation and Prospects* (The Hague: Kluwer Law International, 1998), pp.17–36, quote on p.18.
12. On this and the subsequent developments in chemical warfare during World War I see Martinez, *Vom Giftpfeil zum Chemiewaffenverbot*, pp.55–121; Crone, *Banning Chemical Weapons*, pp.16–19; SIPRI, *The Problem of Chemical and Biological Warfare. Volume I: The Rise of CB Weapons* (Stockholm: Almqvist and Wiksell, 1971), pp.26–58.
13. Crone, *Banning Chemical Weapons*, p.20.
14. See SIPRI, *The Problem of Chemical and Biological Warfare. Volume IV: CB Disarmament Negotiations, 1920–1970* (Stockholm: Almqvist and Wiksell, 1971), pp.36, 58–71.
15. See SIPRI, *The Rise of CB Weapons*, pp.142–6 and Martinetz, *Vom Giftpfeil zum Chemiewaffenverbot*, pp.160–5.
16. http://www.nti.org/e_research/el_japan_1.html, last accessed 25 November 2003.
17. Martinetz, *Vom Giftpfeil zum Chemiewaffenverbot*, pp.150–3.
18. Crone, *Banning Chemical Weapons*, p.28.
19. See SIPRI, *The Rise of CB Weapons*, especially chapter 5, 'The Non-use of CB Weapons during World War II', pp.329–35.
20. See SIPRI, *The Rise of CB Weapons*, pp.71–5; Martinetz, *Vom Giftpfeil zum Chemiewaffenverbot*, pp.202f.
21. On the Egyptian CW use see SIPRI, *The Rise of CB Weapons*, pp.159–61 and pp.326–41, on US use of tear gases and defoliants ibid., pp.162–210.
22. See Chapter 3 for more details on the BWC.
23. Martinetz, *Vom Giftpfeil zum Chemiewaffenverbot*, pp.207–13.
24. See G. S. Pearson and R. S. Magee, 'Critical evaluation of proven chemical weapon destruction technologies (IUPAC Technical Report)', in *Pure and Applied Chemistry*, 74 (2), 2002, 187–316, quote on 202.
25. See Crone, *Banning Chemical Weapons*, p.95.
26. See paragraph 13 in Part IV (A) of the CWC, at http://www.opcw.org/html/db/cwc/eng/cwc_frameset.html.
27. See Pearson and Magee, 'Critical evaluation', p.213f.
28. Ibid. p.212.

29. See Pearson and Magee, 'Critical evaluation', p.204.
30. Idem.
31. The US pursuit of alternative destruction technologies is conducted by the US Army Chemical Materials Agency's 'Alternative Technologies and Approaches Project' (ATAP), see http://www.cma.army.mil/atap.aspx, last accessed 6 September 2005.
32. D. Noltkamper and S. Burgher, 'Toxicity, Phosgene', article updated 12 February 2003, available at http://www.emedicine.com/emerg/topic849.htm, accessed 18 March 2003.
33. US Department of Labor, Occupational Health and Safety Administration website, last updated 27 April 1999, available at http://www.osha-slc.gov/SLTC/healthguidelines/hydrogencyanide/recognition.html, last accessed 18 March 2003.
34. Ibid.
35. S. Borman, 'Combinatorial chemistry. Researchers continue to refine techniques for identifying potential drugs in "libraries" of small organic molecules', *Chemical and Engineering News*, 24 February 1997, available at http://pubs.acs.org/hotartcl/cenear/972024/comb.html, last accessed 18 November 2003.
36. Ibid.
37. For example the *Journal of Combinatorial Chemistry*, published since 1999 by the American Chemical Society, *Combinatorial Chemistry – an online journal*, published since 1999 by Elsevier Science Publisher, available at http://www.sciencedirect.com/science/journal/14643383, *Combinatorial Chemistry and High-Throughput Screening*, published since 1998 by Bentham Publishing.
38. See M. L. Wheelis, 'Biotechnology and biochemical weapons', *The Nonproliferation Review*, 9 (1), 2002, 48–53.
39. L. J. Browne et al., 'Chemogenomics. A novel information tool for drug discovery', *Pharmaceutical Technology*, 2002, 84ff.
40. J. Walter, 'The ins and outs of data mining', in *R&D Magazine*, 45 (4), April 2003, 33f. P. A. Whittaker, 'What is the relevance of bioinformatics to pharmacology?', *Trends in Pharmacological Sciences*, 24 (8), August 2003, 34–9.
41. See A. F. Cowman and B. S. Crabb, 'Functional genomics: identifying drug targets for parasitic diseases', *Trends in Parasitology*, 19 (11), November 2003, 538–43.
42. See G. Vogt, 'Multi-axis robots bring automation to life sciences', *Industrial Robot: An International Journal*, 29 (1), 49–52.
43. See J. Holland, and T. Mitchel, 'Chemists harness IT to organize data and optimize leads', *R&D Magazine*, 41 (10), September 1999, 23f.
44. S. K. Sahoo, and V. Labhasetwar, 'Nanotech approaches to drug delivery and imaging', *Drug Discovery Today*, 8 (24), December 2003, 1112–20; S. S. Davis, 'Biomedical applications of nanotechnology – implications for drug targeting and gene therapy', *Trends in Biotechnology*, 15, June 1997, 217–24.
45. A. Wood, and A. Scott, 'Combinatorial chemistry picks up speed', *Chemical Week*, 162 (30), 9 August 2000, 39–42.

46. R. Pawlicak and J. H. Shelhamer, 'Application of functional genomics in allergy and clinical immunology', *Allergy*, 58, 2003, 973–80.
47. B. Gaston, 'Functional genomics and proteomics in control of breathing', *Respiratory Physiology & Neurobiology*, 135, 2003, 231–38.
48. M. Mackiewicz, and A. I. Pack, 'Functional genomics of sleep', *Respiratory Physiology & Neurobiology*, 135, 2003, 207–20.
49. M. Yamada, and T. Higuchi, 'Functional genomics and depression research. Beyond the monoamine hypothesis', *European Neuropsychopharmacology*, 12, 2002, 235–44.
50. See for example the webpage of the Synthetic Organic Chemical Manufacturers Association at http://www.socma.org/issues/batch.htm, the majority of whose members follow this batch-production approach.
51. See the info brochure of Hoechst industrial park, available at http://www.industriepark-hoechst.com/standortfolder_englisch_.pdf, last accessed 25 November 2003.
52. See Center for Nonproliferation Studies, *The Moscow Theater Hostage Crisis: Incapacitants and Chemical Warfare*, available at http://cns.miis.edu/pubs/week/02110b.htm, last accessed 6 September 2005.
53. See the website of the Sunshine Project for a documentation of the US non-lethal weapons programmes at www.sunshine-project.org.
54. See R. F. Bellamy, 'Medical effects of conventional weapons', *World Journal of Surgery*, 16, 1992, 888–92, quoted in L. Klotz, M. Furmanski and M. Wheelis, *Beware the Siren's Song: Why 'Non-Lethal' Incapacitating Chemical Agents Are Lethal*, April 2003, available at the Federation of American Scientists website www.fas.org/bwc/papers/sirens_song.pdf, last accessed 28 November 2003.
55. See SIPRI, *The Problem of Chemical and Biological Warfare. Volume I: The Rise of CB Weapons* (Stockholm: Almqvist & Wiksell, 1971), p.129.
56. The tear gas CS has a much higher safety margin compared to the incapacitating chemicals under investigation by various militaries today. CS therefore qualifies as a 'riot control agent' under the CWC.
57. 'Non-Lethal' Weapons, the CWC and the BWC, Editorial, *The CBW Conventions Bulletin*, 61, September 2003, 2.
58. The tasks of PTS and PrepCom were spelled out in the so-called Paris Resolution, which was adopted in January 1993 when the CWC was opened for signature.
59. R. Trapp, *Verification under the Chemical Weapons Convention: On-Site Inspection in Chemical Industry Facilities*, SIPRI Chemical and Biological Warfare Studies, No.14 (Oxford: Oxford University Press, 1993), p.8.
60. See Art. XV, paras. 4 and 5 of the CWC for the streamlined amendment procedure of the Annexes to the Convention.
61. Introductory text for the '2nd European Symposium on Non-Lethal Weapons', which was organized by the Fraunhofer Institut Chemische Technologie on 13 and 14 May 2003 in Ettlingen, Germany; available at http://www.ict.fhg.de/english/events/nlw.html, last accessed 31 August 2005.

62. Since 1996 the Joint Non-Lethal Weapons Directorate of the US Marines coordinates the US efforts to develop new non-lethal weapons; see http://www.jnlwd.usmc.mil/, last accessed 28 November 2003. For an assessment of programmes and proposals to streamline and redirect these see National Research Council of the National Academies: *An Assessment of Non-Lethal Weapons Science and Technology* (Washington, DC: The National Academies Press, 2003).
63. B. H. Rosenberg, 'Riot Control Agents and the Chemical Weapons Convention', paper submitted to the 'Open Forum on Challenges to the Chemical Weapons Ban', The Peace Palace, The Hague, 1 May 2003, available at: http://www.fas.org/bwc/papers/rca.pdf, last accessed 28 November 2003.
64. South Korea still insists that it not be named as CW possessor in any official OPCW document. Yet, although the country appears only in coded language as 'a state party' in the OPCW context, it is an open secret that this reference applies to the Republic of Korea.
65. Pamela Mills, 'Progress in The Hague: Quarterly Review no.35', *CBW Conventions Bulletin,* 53, September 2001, 13.
66. Bosnia and Herzegovina, China, France, India, Iran, Japan, Russia, the Federal Republic of Yugoslavia and South Korea.
67. Belgium, Canada, France, Germany, Italy, Japan, Slovenia, the United Kingdom and the United States.
68. China, Italy and Panama.
69. See OPCW, *Annual Report 2000.*
70. See *CBW Conventions Bulletin,* no.46, December 1999, p.13.
71. See A. Kelle, 'The CWC after its first review conference: is the glass half full or half empty?', *Disarmament Diplomacy,* 71, June/July 2003, 31–40.
72. R. J. Mathews, 'Intention of Article VI: an Australian Drafter's Perspective', *OPCW Synthesis,* November 2000, available at http://www.opcw.org/synthesis, last accessed 30 March 2004.
73. Trapp, *Verification under the Chemical Weapons Convention,* p.10
74. *Note by the Director General,* OPCW document RC-1/DG.1, p.12.
75. *Statement to the First Special Session of the Conference of States Parties to Review the Operation of the Chemical Weapons Convention by Mr Mustafa Kamal Kazi, Ambassador and Permanent Representative of Pakistan to the OPCW,* The Hague, 30 April 2003.
76. See A. Kelle, 'Business as usual in implementing the CWC? Not quite yet!', in *Disarmament Diplomacy,* 32, November 2001, 8–12.
77. See *Decision. Provisions on Transfers of Schedule 3 Chemicals to States Not Party to the Convention,* OPCW Document C-VI/DEC.10, The Hague, 17 May 2001.
78. See IUPAC, *Impact of Scientific Developments on the Chemical Weapons Convention. Report by the International Union of Pure and Applied Chemistry to the Organization for the Prohibition of Chemical Weapons,* available on the IUPAC-website at www.iupac.org/publications/pac/2002/pdf/7412x2323.pdf, also published in *Pure and Applied Chemistry,* 74 (12), 2002, 2323–52.

79. *Note by the Director General. Report of the Scientific Advisory Board on Developments in Science and Technology*, OPCW document RC-1/DG.2, The Hague, 23 April 2003, p.15.
80. See http://www.sussex.ac.uk/Units/spru/hsp/OpenForumCWC.pdf, which contains all contributions to the Open Forum, last accessed 6 September 2005.
81. See para 7.21 of the *Report of the First Special Session of the Conference of States Parties to Review the Operation of the Chemical Weapons Convention (First Review Conference) 28 April–9 May 2003*, OPCW document RC-1/5, p.7, available at http://www.opcw.org/docs/rc105.pdf, last accessed 6 September 2005.

3 Science, Technology and the BW Prohibition Regime

1. See M. L. Wheelis, 'Biological Warfare Before 1914', in E. Geissler and J. E. van Courtland Moon (eds), *Biological and Toxin Weapons: Research, Development and Use from the Middle Ages to 1945*, SIPRI Chemical & Biological Warfare Studies, No.18 (Oxford: Oxford University Press, 1999), 8–34.
2. There is indeed an area of overlap in the category of toxins which fall under the regulations of both the BTW convention and the CW convention, the latter of which lists two toxins – ricin and saxitoxin – on its Schedules of CW agents; for more detail see Chapter 4.
3. See Dean Wilkening, 'BCW Attack Scenarios', Sidney D. Drell, Abraham D. Sofaer and George D. Wilson (eds), *The New Terror. Facing the Threat of Biological and Chemical Weapons* (Stanford, CA: Hoover Institution Press, 1999), 76–114.
4. For up to date information on this topic see the Federation of American Scientists website on agricultural biowarfare and bioterrorism at http:// fas. org/bwc/agr/main.htm, maintained by Mark Wheelis, University of California, Davis.
5. Malcolm Dando, 'The impact of the development of modern biology and medicine on the evolution of offensive biological warfare programs in the twentieth century', in *Defense Analysis*, 15 (1), 1999, 43–69, quotes from 51.
6. See Jonathan Tucker, 'Biological weapons in the former Soviet Union: an interview with Dr Kenneth Alibek', *The Nonproliferation Review*, 6 (3), Spring-Summer 1999, 1–10, quote from p.2.
7. Sally Smith Hughes, 'Making dollars out of DNA. The first major patent in biotechnology and the commercialization of molecular biology, 1974–1980', *Isis*, 92, 2001, 541–75, quote on 541.
8. See Stephanie Jones, *The Biotechnologists and the Evolution of Biotech Enterprises in the USA and Europe* (Basingstoke: Macmillan – now Palgrave Macmillan, 1992).
9. Ernst & Young, *Beyond Borders. The Global Biotechnology Report 2002*, p.11.

10. See section 3.2.1. for a discussion of recent developments.
11. Central Intelligence Agency, Directorate of Intelligence, *The Darker Bioweapons Future*, unclassified, Washington, DC, 3 November 2003, p.1f., last accessed on 10 March 2004 at www.fas.org/irp/cia/product/bw1103.pdf.
12. Ernst & Young, *Beyond Borders 2002*, 11f.
13. The official title of the act, as amended by the US Congress is *A bill to amend the Public Health Service Act to provide protections and countermeasures against chemical, radiological, or nuclear agents that may be used in a terrorist attack against the United States by giving the National Institutes of Health contracting flexibility, infrastructure improvements, and expediting the scientific peer review process, and streamlining the Food and Drug Administration approval process of countermeasures*; S.15 was passed by the Senate on 19 May 2004 and after it was agreed upon by the House of Representatives as well, it became Public Law No: 108-276 on 21 July 2004.
14. Henceforth called the '1925 Geneva Protocol', original text in *League of Nations Treaty Series*, Vol. 94, available on numerous websites.
15. On the negotiations see SIPRI, *The Problem of Chemical and Biological Warfare, Vol.IV: CB Disarmament Negotiations, 1920–1970* (Stockholm: Almquist & Wiksel, 1971), pp.58–71.
16. See the list of states parties at http://projects.sipri.se/cbw/docs/cbw-hist-geneva-parties.html.
17. See R. R. Baxter and T. Buergenthal, 'Legal aspects of the Geneva Protocol of 1925', *American Journal of International Law*, 64 (4), 1970, 853–79.
18. According to Sims 12 states have withdrawn their Geneva Protocol reservations since 1986; see N. A. Sims, *The Evolution of Biological Disarmament*, SIPRI Chemical and Biological Warfare Studies, No.19 (Oxford: Oxford University Press, 2001), 152–62.
19. See for example the Soviet proposal submitted to the Eighteen Nation Disarmament Committee in March 1962, as quoted in SIPRI, *The Problem of Chemical and Biological Warfare, Volume IV: CB Disarmament Negotiations, 1920–1970* (Stockholm: Almquist & Wiksel, 1971), pp.231ff.
20. See SIPRI, *CB Disarmament Negotiations*, 254ff; on the development of this working paper see S. Wright, 'Geopolitical Origins', in S. Wright (ed.), *Biological Warfare and Disarmament. New Problems / New Perspectives* (Lanham: Rowman & Littlefield, 2002), 322–33.
21. See J. B. Tucker, 'A farewell to germs. The US renunciation of biological and toxin warfare, 1969–70', *International Security*, 27 (1), 2002, 107–48.
22. See Nicholas A. Sims, *The Diplomacy of Biological Disarmament. Vicissitudes of a Treaty in Force, 1975–85* (New York: St. Martin's Press – now Palgrave Macmillan, 1988), 52–4.
23. See Sims, *The Diplomacy of Biological Disarmament*, 155–9 and 226–52.
24. *Final Declaration of the First Review Conference*, document BWC/CONF.I/10, p.2, available at http://www.opbw.org/rev_cons/1rc/docs/final_dec/1RC Final Doc.pdf, last accessed 12 July 2004.
25. *Final Document of the Second Review Conference*, document BWC/CONF.II/13/II, p.3, available at http://www.opbw.org/rev_cons/2rc/docs/final_dec/2RC Final Doc.pdf, last accessed 12 July 2004.

26. Final Document of the Fourth Review Conference, document BWC/CONF.IV/9, available at http://www.opbw.org/rev_cons/4rc/docs/final_dec/4RC_final_dec.pdf, last accessed 12 July 2004.
27. The US submission to the Review Conference, as well as all other national assessments, is contained in Background Paper on New Scientific and Technological Developments Relevant to the Convention on the Production of the Development, Production and Stockpiling of Bacteriological (Biological) and Toxin Weapons and on their Destruction, document BWC/CONF.V/4, 13–22, quote from p.13, available at http://www.opbw.org/rev_cons/5rc/docs/rev_con_docs/i_docs/V-04.pdf.
28. Ibid., p.2.
29. See section 5.1. on the Fifth Review Conference in general.
30. *Final Declaration of the First Review Conference*, document BWC/CONF.I/9, p.9.
31. The paper by Robert Mikulak is quoted in Nicholas A. Sims, *The Evolution of Biological Disarmament*, SIPRI Chemical and Biological Warfare Studies, No.19 (Oxford: Oxford University Press, 2001), p.63.
32. I. Hunger, 'Article V: Confidence Building Measures', in G. Pearson and M. Dando (eds), *Strengthening the Biological Weapons Convention. Key Points for the Fourth Review Conference* (Geneva: QUNO, 1996), pp.77–92, quote from p.78; see also M. I. Chevrier and I. Hunger, 'Confidence-building measures for the BTWC: performance and potential', *The Nonproliferation Review*, 7 (3), 2000, 24–42.
33. B. Roberts, 'Export controls and biological weapons: new roles, new challenges', *Critical Reviews in Microbiology*, 24 (3), 1998, 235–54.
34. 'Introduction', available at http://www.australiagroup.net/en/intro.htm, last accessed 14 July 2004.
35. See J. P. Robinson, 'The Australia Group: a Description and Assessment', in H. G. Brauch et al. (eds), *Controlling the Development and Spread of Military Technology. Lessons From the Past and Challenges for the Future* (Amsterdam: VU University Press, 1992), pp.157–76.
36. *Media Release. 2004 Australia Group Plenary*, available at http://www.australiagroup.net/en/releases/press_2004_06.htm, last accessed 14 July 2004.
37. Ad-Hoc Group of Governmental Experts to Identify and Examine Potential Verification Measures From a Scientific and Technical Standpoint, *Report*, Geneva, 1993, document BWC/CONF.III/VEREX/9, 8.
38. Sims, *The Evolution of Biological Disarmament*, p.104; the mandate of the AHG is contained in Special Conference of the States Parties to the Convention on the Prohibition of the Development, Production and Stockpiling of Bacteriological (Biological) and Toxin Weapons and on their Destruction (Geneva, 19–30 September 1994), *Final Report*, document BWC/SPCONF/1, 10f.
39. 'Testimony by Ambassador Donald Mahley, State Department Special Negotiator for Chemical and Biological Arms Control before the House Committee on Government Reform, Subcommittee on National Security, Veterans' Affairs and International Relations, September 13', reprinted

partially in *Disarmament Diplomacy*, 50, September 2000, 37–40, quote on p.39.

40. The requirements for effective verification were spelt out for example by Ambassador Mahley during the 1994 Special Conference of the States Parties to the BWC. See *Statement of US Representative Donald A. Mahley to the Committee of the Whole, September 22, 1994*, document BWC/SPCONF/WP.16, last accessed 10 March 2004 at www.opbw.org.
41. On this latter point see T. Bernauer, 'Verification of Compliance with the Biological Weapons Convention: Developing Countries Between Passive Participation and Obstruction', in Oliver Thränert (ed.), *The Verification of the Biological Weapons Convention: Problems and Perspectives*, Report No.50 (Bonn: Friedrich Ebert Foundation, 1992), pp.55–67.
42. The text of the mandate is contained in the Final Report of the Special Conference, see note 38.
43. This characterization follows T. Toth, 'Time to wrap up', in *The Chemical and Biological Weapons Conventions Bulletin*, 46, 1999, 1–3.
44. J. Rissanen, 'Chair releases his "composite text" for verification protocol', in *Disarmament Diplomacy*, 55, London: Acronym Institute, March 2001. See also the assessment given by G. S. Pearson, M. R. Dando and N. A. Sims, *The Composite Protocol Text: an Effective Strengthening of the Biological and Toxin Weapons Convention*, Evaluation Paper No.21 (Bradford: University of Bradford, July 2001).
45. See V. Beck, 'Implications for Biological Defence of Legally Binding Declarations and Declaration Follow-Up Procedures', in M. I. Chevrier et al. (eds), *The Implementation of Legally Binding Measures to Strengthen the Biological and Toxin Weapons Convention*, NATO Science Series II/150 (Dordrecht: Kluwer Academic Publishers, 2004), pp.139–44.
46. See Working Paper Submitted by Germany and Sweden, *Proposed Language for Article III – Declarations*, presented to the thirteenth session of the AHG, Geneva, 7 January 1999, Document BWC/AD HOC GROUP/WP.340.
47. On the following see the section 'Article 4 Declarations', in BWC/AD HOC GROUP/56–2, 364–9.
48. For a more detailed discussion of the pros and cons of non-challenge visits see D. MacEachin, 'Routine and challenge: two pillars of verification', *The CBW Conventions Bulletin*, 39, March 1998, 1–3.
49. See H. Wilson, 'BWC update', *Disarmament Diplomacy*, 42, December 1999, 27–34.
50. See H. Wilson, 'BWC Update'.
51. On the following see 'Article 9 Investigations', in BWC/AD HOC GROUP/56–2, pp.404–14, quote from p.404
52. On the latter see M. Wheelis, 'Investigating disease outbreaks under a Protocol to the Biological and Toxin Weapons Convention', in *Emerging Infectious Diseases*, 6 (6), 2000, 595–600.
53. See *Working Paper by India – Guidelines to ensure Compliance with Obligations under Article III of the Convention on the Prohibition …*, Document BWC/AD HOC GROUP/WP.126, 5 March 1997.

54. See *Working Paper by Austria and New Zealand*, BWC/AD HOC GROUP/WP.142, 14 March 1997.
55. See Working paper submitted by the Group of NAM and Other Countries, *Establishment of a Cooperation Committee*, BWC/AD HOC GROUP/WP.349, January 1999.
56. See Jenni Rissanen, 'The BWC Protocol negotiation 18th session: removing brackets', *Disarmament Diplomacy*, 43, January/February 2000, 21–5.
57. Article 7, Paragraph 1, BWC/Ad HOC GROUP/56–2, 396
58. See Parargaph 5 of Article 7 of the Protocol text, as contained in BWC/AD HOC GROUP/56–2, 397.
59. See Section D of Article 14, in BWC/AD HOC GROUP/56–2, 427–32.
60. See *Final Document* of the Fourth Review Conference, document BWC/CONF.IV/9, p.27.
61. M. R. Dando, *Preventing Biological Warfare: the Failure of American Leadership* (Basingstoke: Palgrave – now Palgrave Macmillan, 2002), p.175.
62. B. H. Rosenberg, 'Allergic reaction: Washington's response to the BWC Protocol', *Arms Control Today*, July/August 2001, 3–8, available at http://www.armscontrol.org/act/2001_07-08/rosenbergjul_aug01.asp
63. See G. S. Pearson, M. R. Dando and N. A. Sims, *The US Rejection of the Composite Protocol: a Huge Mistake based on Illogical Assessments*, Evaluation Paper No.22, Bradford, August 2001, available at http://www.brad.ac.uk/acad/sbtwc/evaluation/evalu22.pdf.
64. J. Rissanen, *Acrimonious Opening for BWC Review Conference*, BWC Rev.Con. Bulletin November 19, 2001, available at www.acronym.org.uk/bwc/index.htm.
65. The draft final document of the Conference as issued by the Chairman of the Committee of the Whole is available at www.acronym.org.uk/bwc/index.htm.
66. See J. Rissanen, 'Left in Limbo: Review Conference suspended on edge of collapse', *Disarmament Diplomacy*, 62, January–February 2002, 18–32.
67. See the detailed analysis in M. I. Chevrier, 'Waiting for Godot or saving the show? The BWC Review Conference reaches modest agreement', *Disarmament Diplomacy*, 68, December 2002–January 2003, 11–16.
68. Final Document of the Fifth BWC Review Conference, document BWC/CONF.V/17, p.3, available at http://disarmament2.un.org/wmd/bwc/pdf/bwccnfv17.PDF.
69. See J. Littlewood, 'Substance hidden under a mountain of paper: the BWC experts meeting in 2003', *Disarmament Diplomacy*, 73, October–November 2003, 63–6.
70. Ibid.
71. J. B. Tucker, 'The BWC new process: a preliminary assessment', *The Nonproliferation Review*, 11 (1), Spring 2004, 26–38, quote from p.32.
72. G. S. Pearson, 'The Biological Weapons Convention Meeting of states parties', *The CBW Conventions Bulletin*, 66, December 2004, pp.21–34, quote on p.33.

73. See the summaries of these cases in: National Research Council of the National Academies, Committee on Research Standards and Practices to Prevent the Destructive Application of Biotechnology, *Biotechnology Research in an Age of Terrorism* (Washington, DC: The National Academies Press, 2004), pp.24–9.
74. NAS, *Biotechnology Research in an Age of Terrorism*, p.27.
75. Ibid, p.32.
76. Ibid., pp.4–12.
77. See the NSABB's website at http://www.biosecurityboard.gov/.
78. The Secretary of Health and Human Services, *Charter. National Science Advisory Board for Biosecurity*, Washington, DC, 4 March 2004, available at http://www.biosecurityboard.gov/SIGNED%20NSABB%20Charter.pdf.
79. NSABB Charter, p.2.
80. See Journal Editors and Authors Group, 'Statement on scientific publication and security', in *Science*, 299 (5610), 1149, available at http://www.sciencemag.org/cgi/reprint/299/5610/1149.pdf.
81. Statement on Scientific Publication and Security, as reprinted in NAS, *Biotechnology Research in an Age of Terrorism*, 98–9.

4 Defences Under Attack: the Potential Misuse of Immunology

1. R. Nowak, 'Disaster in the making. An engineered mouse virus leaves us one step away from the ultimate bioweapon', *New Scientist,* 13 January 2001, 4–5; R. J. Jackson et al., 'Expression of mouse interleukin-4 by a recombinant ectromelia virus suppresses cytolytic lymphocyte responses and overcomes genetic resistance to mousepox', *Journal of Virology*, 75, 2001, 1205–10.
2. M. Buller, *The potential use of genetic engineering to enhance orthopoxviruses as bioweapons*. Presentation at the International Conference 'Smallpox Biosecurity. Preventing the Unthinkable' (21–22 October 2003) Geneva, Switzerland; J. D. Steinbruner and E. D. Harris, 'When science breeds nightmares', *International Herald Tribune,* 3 December 2003; D. MacKenzie, 'US develops lethal new viruses', *New Scientist*, 180, 6, 2003.
3. A. M. Rosengard et al., 'Variola virus immune evasion design: expression of a highly efficient inhibitor of human complement', *Proceedings of the National Academy of Sciences USA*, 99, 2002, 8808–13.
4. A. P. Pomerantsev, et al., 'Expression of cereolysine AB genes in *Bacillus anthracis* vaccine strain ensures protection against experimental infection', *Vaccine*, 15, 1997, 1846–50.
5. J. E. v. C. Moon, 'The US BW Program: Dilemmas of Policy and Preparedness', in M. Wheelis, L. Rozsa and M. R. Dando (eds), *Deadly Cultures: Bioweapons from 1945 to the Present* (Cambridge, MA: Harvard University Press, 2006), pp. 9–46.
6. E. Geissler and K. Lohs, 'The Changing Status of Toxin Weapons', in E. Geissler (ed.), *Biological and Toxin Weapons Today* (Oxford: Oxford University Press 1986), pp.36–56.

7. NIH, *NIAID biodefense research agenda for CDC category A agents. Progress Report*, August 2003, available at http://www.niaid.nih.gov/biodefense/research/bioresearchagenda.pdf.
8. R. A. Goldsby et al., *Immunology*, 5th edition (New York: W. H. Freeman and Company, 2003); A. K. Abbas, A. H. Lichtman, and J. S. Pober, *Cellular and Molecular Immunology*, 3rd edition (Philadelphia: W. B. Saunders Company, 1997).
9. R. Medzhitov and C. A. Janeway Jr., 'An ancient system of host defense', *Current Opinion in Immunology*, 10, 1998, 12–15.
10. R. Medzhitov, P. Preston-Hurlburt and C. A. Janeway, Jr., 'A human homologue of the *Drosophila* Toll protein signals activation of adaptive immunity', *Nature*, 388, 1997, 394–7; S. Akira, 'Mammalian Toll-like receptors', *Current Opinion in Immunology*, 15, 2003, 5–11; M. Triantafilou and K. Triantafilou, 'Lipopolysaccharide recognition: CD14, TLRs and the LPS-activation cluster', *Trends in Immunology*, 23, 2002, 301–04.
11. E. T. Rietschel and H. Brade, 'Bacterial endotoxins', *Scientific American*, 267, 1992, 54–61.
12. USAMRIID, 'Basic studies seeking generic medical countermeasures against agents of biological origin', in USAMRIID, *Annual Report for Fiscal Year 1987*, p.19; B. Rosenberg and G. Burck, 'Verification of Compliance with the Biological Weapons Convention', in S. Wright (ed.), *Preventing a Biological Arms Race* (Cambridge, MA: MIT Press, 1990), pp. 301–29.
13. J. E. Parker, 'Plant recognition of microbial patterns', *Trends in Plant Science*, 8, 6, 2002, 245–7; T. Nürnberger, and F. Brunner, 'Innate immunity in plants and animals: emerging parallels between the recognition of general elicitors and pathogen-associated molecular patterns', *Current Opinion in Plant Biology*, 5, 2002, 318–24.
14. N. Inohara and G. Nunez, 'Nods: intracellular proteins involved in inflammation and apoptosis', *Nature Reviews Immunology*, 3, 2003, 371–82; D. A. Jones and D. Takemoto, 'Plant innate immunity – direct and indirect recognition of general and specific pathogen-associated molecules', *Current Opinion in Immunology*, 16, 2004, 48–62.
15. J. Cohn, G. Sessa and G. B. Martin, 'Innate immunity in plants', *Current Opinion in Immunology*, 13, 2001, 55–62.
16. Ibid.
17. G. P. Bolwell, 'Role of active oxygen species and NO in plant defence responses', *Current Opinion in Plant Biology*, 2, 1999, 287–94.
18. S. Gupta, N. Ferguson and R. Anderson, 'Chaos, persistence, and evolution of strain structure in antigenically diverse infectious agents', *Science*, 280, 1998, 912–15.
19. See Goldsby et al., *Immunology*, note 8.
20. A. Alcami and U. H. Koszinowski, 'Viral mechanisms of immune evasion', *Trends in Microbiology*, 8, 2000, 410–18; D. Tortorella et al., 'Viral subversion of the immune system', *Annual Review of Immunology*, 18, 2000, 861–926.
21. See Rosengard et al., 2002, note 3.
22. See Alcami and Koszinowski, 2000, note 20.

23. See Alcami and Koszinowski, 2000, note 20.
24. Ibid.
25. L. N. Carayannopoulos and W. M. Yokoyama, 'Recognition of infected cells by natural killer cells', *Current Opinion in Immunology*, 16, 2004, 26–33.
26. M. R. Dando, V. Nathanson and M. Darvell, 'The impact of biotechnology', The British Medical Association (ed.), *Biotechnology, Weapons and Humanity* (Amsterdam: Harwood Academic Publishers), 1999, pp.33–51; K. Nixdorff et al., *Biotechnology and the Biological Weapons Convention* (Münster: Agenda Verlag, 2003).
27. See Pomerantsev et al., 1997, note 4.
28. Nowak, 2001, note 1.
29. R. J. Jackson, et al., 'Infertility in mice induced by a recombinant ectromelia virus expressing mouse zona pellucida glycoprotein', *Biology of Reproduction*, 58, 1998, 152–9; Jackson et al., 2001, note 1.
30. Jackson et al., 1998, note 29.
31. See Jackson et al., 2001, note 1.
32. Ibid.
33. See Buller, 2003, note 2.
34. See MacKenzie, 2003, note 2.
35. Ibid.
36. See Steinbruner and Harris, 2003, note 2.
37. National Research Council of the National Academies, *Biotechnology Research in an Age of Terrorism: Confronting the Dual Use Dilemma* (Washington, DC: The National Academies Press, 2003).
38. T. M. Tumpey et al., 'Existing antivirals are effective against influenza viruses with genes from the 1918 pandemic virus', *Proceedings of the National Academy of Sciences USA*, 99, 2002, 13849–54.
39. A. H. Reid, et al., 'Characterization of the 1918 "spanish" influenza virus neuraminidase gene', *Proceedings of the National Academy of Sciences USA*, 97, 2000, 6785–90.
40. See Tumpey et al., 2002, note 38.
41. National Research Council, 2003, note 37.
42. J. Cello, A. V. Paul and E. Wimmer, 'Chemical synthesis of poliovirus cDNA: generation of infectious virus in the absence of natural template', *Science*, 297, 2002, 1016–18.
43. H. O. Smith et al., 'Generating a synthetic genome by whole genome assembly: ϕX174 bacteriophage from synthetic oligonucleotides', *Proceedings of the National Academy of Sciences USA*, 100 (2003), 15440–45.
44. J. Couzin, 'Bioterrorism: a call for restraint on biological data', *Science*, 297, 2002, 749–51.
45. A. Domi and B. Moss, 'Cloning the vaccinia virus genome as a bacterial artificial chromosome in *Escherichia coli* and recovery of infectious virus in mammalian cells', *Proceedings of the National Academy of Sciences USA*, 99, 2002, 12415–20.
46. A. Inui, 'Cytokines and sickness behaviour: implications from knockout models', *Trends in Immunology*, 22, 2001, 469–73.

47. See Rietschel and Brade, 1992, note 11.
48. See Goldsby et al., 2003, note 8.
49. See Moon, 2006, note 5.
50. See Geissler and Lohs, 1986, note 6.
51. See Goldsby et al., 2003, note 8.
52. G. J. Silverman et al., 'The dual phases of the response to a neonatal exposure to a V_H family-restricted staphylococcal B cell superantigen', *Journal of Immunology*, 161, 1998, 5720–32; C. S. Goodyear and G. J. Silverman, 'Death by a B cell superantigen: in vivo V_H-targeted apoptotic supraclonal B cell deletion by a staphylococcal toxin', *Journal of Experimental Medicine*, 197, 2003, 1125–39.
53. K. Minton, 'Immune evasion. Germ warfare', *Nature Reviews Immunology*, 3, 2003, 442.
54. B. Moss, 'Vaccinia virus expression vector: a new tool for immunologists', *Immunology Today*, 6, 1985, 243–5; J. A. McCart et al., 'Systemic cancer therapy with a tumor-selective vaccinia virus mutant lacking thymidine kinase and vaccinia growth factor', *Cancer Research*, 61, 2001, 8751–7.
55. M. A. Morsy and C. T. Caskey, 'Safe gene vectors made simpler', *Nature Biotechnology*, 15, 1997, 17; S. Kochanek et al., 'A new adenoviral vector: Replacement of all viral coding sequences with 28 kb of DNA independently expressing both full-length dystrophin and β-galactosidase', *Proceedings of the National Academy of Sciences USA*, 93, 1996, 5731–6.
56. B. J. Carter, 'The promise of adeno-associated virus vectors', *Nature Biotechnology*, 14, 1996, 1725–6.
57. E. Check, 'Harmful potential of viral vectors fuels doubts over gene therapy', *Nature*, 423, 2003, 573–4.
58. R. J. Kreitman, 'Immunotoxins in cancer therapy', *Current Opinion in Immunology*, 11, 1999, 570–8.
59. M. S. Hayden, L. K. Gilliland and J. A. Ledbetter, 'Antibody engineering', *Current Opinion in Immunology*, 9, 1997, 201–12.
60. See Goldsby et al., 2003, note 8.
61. See Hayden et al., 1997, note 60.
62. See USAMRIID, 1987; Rosenberg and Burck, 1990, note 12.
63. G. J. V. Nossal, 'Vaccines', in W. E. Paul (ed.), *Fundamental Immunology*, 5th edition (Philadelphia: Lippencott Williams & Wilkins, 2003), pp.1319–69; S. J. Streatfield et al., 'Plant-based vaccines: unique advantages', *Vaccine*, 19, 2001, 2742–8.
64. See Nossal, 2003, note 64; E. Marquet-Blouin et al., 'Neutralizing immunogenicity of transgenic carrot (*Daucus carota* L.)-derived measles virus hemagglutinin', *Plant Molecular Biology*, 51, 2003, 459–69.
65. See Streatfield et al., 2001, note 64.
66. Y. Thanavala et al. 'Immunogenicity of transgenic plant-derived hepatitis B surface antigen', *Proceedings of the National Academy of Sciences USA*, 92, 1995, 3358–61.
67. See Marquet-Blouin et al., 2003, note 65; Streatfield et al., 2003, note 64.

68. N. Matoba et al., 'A mucosally targeted subunit vaccine candidate eliciting HIV-1 transcytosis-blocking Abs', *Proceedings of the National Academy of Sciences USA*, 101, 2004, 13584–9; M. M. Rigano et al., 'Production of a fusion protein consisting of the enterotoxigenic *Escherichia coli* heat-labile toxin B subunit and a tuberculous antigen in *Arabidopsis thaliana*', *Plant Cell Reports*, 22, 2004, 502–08.
69. See Carayannopoulos and Yokoyama, 2004, note 25.
70. See Goldsby et al., 2003, note 8.
71. G. B. E. Stewart-Jones et al., 'A structural basis for immunodominant human T cell receptor recognition', *Nature Immunology*, 4, 2003, 657–63.

5 Behaviour Under Control: the Malign Misuse of Neuroscience

1. SIPRI, *The Problem of Chemical and Biological Warfare*, Volume II of *CB Weapons Today* (Stockholm: Almqvist and Wiksell, 1973), pp.288–308 on novel chemical agents.
2. M. R. Dando, 'Scientific and technological change and the future of the CWC: The problem of non-lethal weapons', *Disarmament Forum*, 4, 2002, 33–44.
3. M. R. Dando, *The Danger to the Chemical Weapons Convention from Incapacitating Chemicals*, CWC Review Conference Paper No. 4, Department of Peace Studies, University of Bradford, 2003, available at: http://www.brad.ac.uk/acad/scwc.
4. H. Kitano, 'Systems biology: a brief overview', Introduction to a special section on systems biology: the Genome, Ligome and Beyond, *Science*, 295, 2002, 1662–4.
5. E. J. Davidov, et al., 'Advancing drug discovery through systems biology', *Drug Discovery Today*, 8 (4), 2003, 175–83.
6. C. M. Henry, 'Systems biology: Integrative approach to drug discovery', *Chemical and Engineering News*, 19 May 2003, 45–55.
7. L. Hood, 'Leroy Hood expounds the principles, practice and future of systems biology', *Drug Discovery Today*, 8 (10), 2003, 436–8.
8. M. R. Dando, *A New Form of Warfare: the Rise of Non-Lethal Weapons* (London: Brassey's, 1996), especially pp.136–68 (chapter 8, 'An Assault on the Brain').
9. J. M. Lakoski, et al., *The Advantages and Limitations of Calmatives for Use as a Non-Lethal Technique*, Applied Research Laboratory, College of Medicine, Pennsylvania State University, 2000.
10. A. Longstaff, *Instant Notes: Neuroscience*, 2nd edition (New York: Taylor and Francis, 2005).
11. Editorial, 'New vistas for an old neurotransmitter', *Biological Psychiatry*, 46 (9), 1999, 1121–3. (Introduction to a special issue covering pages 1121–320.)
12. G. Aston-Jones, et al., 'Role of locus coeruleus in attention and behavioural flexibility', *Biological Psychiatry*, 46 (9), 1999, 1309–20.

13. B. Fernandez-Pastor, and J. J. Meana, 'In vivo tonic modulation of the noradrenaline release in the rat cortex by locus coeruleus somatodendritic alpha2 adrenoceptors', *European Journal of Pharmacology*, 442, 2002, 225–9.
14. Edgewood RDEC, *Scientific Conference in Chemical and Biological Defense Research: Abstract Digest*, US Army Chemical and Biological Defense Command, Aberdeen Proving Ground, Maryland, 1989–94.
15. See Dando, *A New Form of Warfare*, note 8.
16. J. R. Docherty, 'Subtypes of functional α_1- and α_2-adrenoceptors', *European Journal of Pharmacology*, 361 (1), 1998, 1–15.
17. M. M. Bücheler et al., 'Two α_2-adrenergic receptor subtypes, α_{2A} and α_{2C}, inhibit transmitter release in the brain of gene-targeted mice', *Neuroscience*, 109 (4), 2002, 819–26.
18. K. M. Small et al. 'Pharmacology and physiology of human adrenergic receptor polymorphisms', *Annu. Rev. Pharmacol. Toxicol.*, 43, 2003, 381–411.
19. See Dando, *The Danger to the Chemical Weapons Convention*, note 3.
20. SIPRI, *The Problem of Chemical and Biological Warfare*, note 1.
21. A. Frances and M. B. First, *Your Mental Health: a Layman's Guide to the Psychiatrists' Bible* (New York: Scribner, 1998), especially pp.109–16 (chapter 5, 'Exposure to Traumatic Events').
22. Longstaff, *Instant Notes: Neuroscience*, note 10.
23. J. LeDoux, 'The power of emotions', in R. Conlon (ed.), *States of Mind: New Discoveries About How Our Brains Make Us Who We Are* (New York: John Wiley and Sons 1999), pp.123–50.
24. J. L. McGough et al., 'Amygdala modulation of memory consolidation: Interaction with other brain systems', *Neurobiology of Learning and Memory*, 78, 2002, 539–52.
25. B. Ferry and J. L. McGough, 'Role of amygdala norepinephrine in mediating stress hormone regulation of memory storage' *Acta Pharmacol. Sin.*, 21 (6), 2000, 481–93.
26. J. L. McGough and B. Roozendaal, 'Role of adrenal stress hormones in forming lasting memories in the brain', *Current Opinion in Neurobiology*, 12, 2002, 205–10.
27. B. Ferry et al., 'Basolateral amygdala noradrenergic influences on memory storage are mediated by an interaction between β- and α_1-adrenoceptors', *Journal of Neuroscience*, 19 (2), 1999, 5119–123; B. Roozendaal, 'Glucocortcoids and the regulation of memory consolidation', *Psychoneuroendocrinology*, 25, 2000, 213–18.
28. E. Vermetten and J. D. Bremner, 'Circuits and systems in stress: I Preclinical Studies', *Depression and Anxiety*, 15, 2002, 126–47.
29. R. Grossman et al., 'Neuroimaging studied on post-traumatic stress disorder', *Psychiatr. Clin. N. America*, 25, 2002, 317–40.
30. S. M. Southwick et al., 'Role of norepinephrine in the pathophysiology and treatment of post-tramumatic stress disorder', *Biological Psychiatry*, 46 (9), 1999, 1192–204.

31. E. Baard, 'The guilt-free soldier: New science raises the spectre of a world without regret', *The Village Voice*, 22–8 January 2003.
32. F. C. Conahan, *Human Experimentation: an Overview on Cold War Era Programs*, Testimony before the Legislative and National Security Subcommittee, Committee on Government Operations, House of Representatives, September 28th. GAO/T-NSIAD-94-266 (Washington, DC: General Accounting Office 1994).
33. SIPRI, *The Problem of Chemical and Biological Warfare*, pp.302–03.
34. US Army Intelligence Agency, *Letter Report: Incapacitating Agents, European Communist Countries*, AST-1620R-100-90, US Army Foreign Science and Technology Center, 16 July 1990.
35. J. S. Ketchum and F. R. Sidell, 'Incapacitating Agents', in F. R. Sidell, E. T. Takafuji and D. R. Franz (eds), *Medical Aspects of Chemical and Biological Warfare* (Washington, DC: Office of the Surgeon General, 1997), pp.287–306.
36. See M. R. Dando, *The Danger of the Chemical Weapons Convention from Incapacitating Chemicals*.
37. M. R. Dando, *A New Form of Warfare: the Rise of Non-Lethal Weapons* (London: Brassey's, 1996), pp.136–68.
38. E. Kagan, 'Bioregulators as instruments of terror', *Clinics in Laboratory Medicine*, 21 (3), 2001, 607–18.
39. B. Knickerbocker, 'Military looks to drugs for battle readiness: As combat flights get longer, pilot use of amphetamines grows, as do side effects', *The Christian Science Monitor*, 9 August 2002.
40. D-J. Dyk and S. W. Lockley, 'Functional genomics of sleep and circadian rhythm', *J. Appl. Physiol.*, 92, 2002, 852–62; G. K. Wang and A. Sehgal, 'Signalling components that drive circadian rhythms', *Current Opinion in Neurobiology*, 12, 2002, 331–8; R. N. Van Gelder et al., 'Circadian rhythms: in the loop at last', *Science*, 300, 6 June 2003, 152–3.
41. M. R. Dando, *The New Biological Weapons: Threat, Proliferation and Control* (Boulder, CO: Lynne Rienner, 2001), pp.103–16.
42. 'Scope/Information', *Nature Reviews: Neuroscience*, available at: http://www.nature.com/nrn/info/scope.html.
43. T. Ideker, T. Galitski and L. Hood, 'A new approach to decoding life: Systems biology', *Ann. Rev. Genomics Hum. Genet.*, 2, 2001, 343–72.
44. M. M. Francis, J. E. Mellem and A. Villu Mericq, 'Bridging the gap between genes and behaviour: recent advances in the electophysiological analysis of neural function', *Caenorhabditis elegans. Trends in Neuroscience*, 26 (2), 2003, 90–9.
45. United Kingdom, *Background Paper on New Scientific and Technological Developments*. Fifth Review Conference of the States Parties to the Convention on the Prohibition of the Development, Production and Stockpiling of Bacteriological (Biological) and Toxin Weapons and on their Destruction. BWC/CONF.V/4/Add.1 (Geneva, 26 October 2001).
46. M. R. Dando, *A New Form of Warfare: the Rise of Non-Lethal Weapons* (London: Brassey's, 1996), especially chapter 6, 'The Human Nervous System'.

47. R. R. Llinás, *I of the Vortex: From Neurons to Self* (Cambridge, MA: MIT Press, 2002).
48. M. W. Dubin, *How the Brain Works* (Oxford: Blackwell, 2002).
49. F. R. Sidell, 'Nerve Agents', in F. R. Sidell, E. T. Takafuji and D. R. Franz (eds), *Medical Aspects of Chemical and Biological Warfare* (Washington, DC: Office of the Surgeon General, US Army, 1997), pp.129–80.
50. J. S. Ketchum, and F. R. Sidell, 'Incapacitating Agents', note 32.
51. E. J. Nestler et al., *Molecular Neuropharmacology: a Foundation for Clinical Neuroscience* (New York: McGraw-Hill Medical, 2001).
52. C. C. Felder et al., 'Therapeutic opportunities for muscarinic receptors in the central nervous system', *J. Medical Chemistry*, 43 (23), 2000, 4333–53.
53. N. J. M. Birdsall, 'Muscarinic acetylcholine receptors', *The IUPHAR Receptor Compendium* (London: IUPHAR Media, 1998), pp.37–45.
54. F. P. Bymaster et al., 'Use of M1-M5 muscarinic receptor knockout mice as novel tools to delineate the physiological roles of the muscarinic choliergic system', *Neurochem. Res.*, 28 (3–4), 2003, 437–42.
55. W. Zhang et al., 'Characterization of central inhibitory muscarinic autoreceptors by the use of muscarinic acetylcholine receptor knock-out mice', *J. Neuroscience*, 22 (5), 2002, 1709–17.
56. J. E. Lachowicz et al., 'Discovery of SCH211803, a high affinity, selective M2 receptor antagonist and a novel approach to treatment of Alzheimer's Disease', *Soc. Neurosci. Abstr.*, 27, 2001, 679.
57. Y. Wang, et al., 'Improving the oral efficacy of CNS drug candidates: Discovery of highly orally efficacious piperidinyl piperidine M_2 muscarinic receptor antagonists', *J. Medical Chemistry*, 45 (25), 2002, 5415–18.
58. A. Lonsgstaff, *Instant Notes: Neuroscience*. 2nd edition (New York: Taylor and Francis, 2000), especially section M4, 'Brain Biological Clocks'.
59. H. Okamura, 'Integration of mammalian circadian clock signals: from molecule to behaviour', *Journal of Endocrinology*, 177, 2003, 3–6.
60. H. D. Piggins and D. J. Cutler, 'The roles of vasocactive intestinal polypeptide in the mammalian circadian clock', *Journal of Endocrinology*, 177, 2003, 7–15.
61. A. Longstaff, 'Sleep', Section 04, in note 58.
62. W. McDowell Anderson, 'Top ten list in sleep', *Chest*, 122 (4), 2002, 1457–60.
63. S. Taheri et al., 'The role of hypocretins (orexins) in sleep regulation and narcolepsy', *Annu. Rev. Neurosci.*, 25, 2002, 283–313.
64. L. de Lecea, et al., 'The hypocretins: hypothalamus-specific peptides with neuroexcitatory activity', *Proc. Nat. Acad. Sci.*, 95, 1998, 322–7.
65. T. Sakurai et al., 'Orexins and orexin receptors: a family of hypothalamic neuropeptides and G protein-coupled receptors that regulate feeding behaviour', *Cell*, 92, 1998, 573–85.
66. P. Bourgin et al., 'Hypocretin-1 modulates rapid eye movement sleep through activation of locus coeruleus neurons', *J. Neurosci.*, 20, 2000, 7760–7.
67. Ibid.

6 Double Assault: Malign Manipulation of the Neuroendocrine-Immune System

1. M. Karwa et al., 'Bioterrorism: Preparing for the impossible or the improbable', *Critical Care Medicine*, 33, 1 (Suppl.), 2005, 575–95.
2. For a an overview of the three lists see: http://www.bt.cdc.gov/agent/agentlist-category.asp.
3. P. J. Osterbauer and M. R. Dobbs, 'Neurobiological weapons', *Neurologic Clinics*, 23, 2005, 599–621.
4. K. M. Prakosh and Y. L. Lo, 'The role of clinical neurophysiology in bioterrorism', *Acta Neurol. Scand.*, 111, 2005, 1–6.
5. C. O. Martin and H. P. Adams, 'Neurological aspects of biological and chemical terrorism: a review for neurologists' *Arch. Neurol.*, 60, 2003, 21–5.
6. Ibid.
7. Ibid.
8. C. A. Janeway and R. Medzhitov, 'Innate immune recognition', *Ann. Rev. Immunol.*, 20, 2002, 197–216.
9. A. M. Rosengard *et al.*, 'Variola virus immune evasion design: Expression of a highly efficient inhibitor of human complement', *Proc. Natl. Acad. Sci.*, 99, 2002, 8808–13.
10. R. G. Ulrich et al., 'Staphylococcal enterotoxin B and related pyrogenic toxins', in F. R. Sidell *et al.* (eds), *Medical Aspects of Chemical and Biological Warfare* (Washington, DC: Office of the Surgeon General, US Army, 1997), pp.621–30.
11. Ibid.
12. N. J. Mantis, 'Vaccines against the category B toxins: Staphylococcal enterotoxin B, epsilon toxin and ricin', *Advanced Drug Delivery Reviews*, 57, 2005, 1424–39.
13. Ibid.
14. G. Pacheco-Lopez et al., 'Behavioural endocrine immune-conditioned response is induced by taste and superantigen pairing', *Neuroscience*, 129, 2004, 555–62.
15. A. W. Kusnecov and Y. Goldfarb, 'Neural and behavioural responses to systemic immunologic stimuli: A consideration of bacterial T cell superantigens', *Curr. Pharm. Des.*, 11, 2005, 1039–46.
16. A. Baum and D. M. Posluszny, 'Health psychology: mapping biobehavioural contributions to health and illness', *Annu. Rev. Psychol.*, 50, 1999, 137–63.
17. J. I. Webster, L. Tonelli and E. M. Sternberg, 'Neuroendocrine regulation of immunity', *Ann. Rev. Immunol.*, 20, 2002, 125–63.
18. E. M. Sternberg, 'The stress response and the regulation of inflammatory disease', *Annals of Internal Medicine*, 117, 1992, 854–66.
19. National Institutes of Health, 'Stress system malfunction could lead to serious, life threatening disease', *NIH Backgrounder* (9 September 2002), available at: http://www.nih.gov/news/pr/sep2002/nichd-09.htm.

20. I.-G. Rojas et al., 'Stress-induced susceptibility to bacterial infection during cutaneous wound healing', *Brain, Behaviour and Immunity*, 16, 2002, 74–84.
21. J. K. Kiecolt-Glaser et al., 'Psychoneuroimmunology: psychological influences on immune function and health', *Journal of Consulting and Clinical Psychology*, 70, 2002, 537–47.
22. Ibid.
23. M. Kubera and M. Maes, 'Serotonin-immune interactions in major depression', in P. Patterson, C. Kordon and Y. Christen (eds), *Neuro-Immune Interactions in Neurologic and Psychiatric Disorders* (Berlin: Springer, 2000), pp.79–87; J. Gordon and N. M. Barnes, 'Lymphocytes transport serotonin and dopamine: agony or ecstasy?', *Trends in Immunology*, 24, 2003, 438–43.
24. National Institutes of Health, 'Gene more than doubles risk of depression following life stresses', *NIH News Release* (17 July 2003), available at: www.nimh.nih.gov/events/prgenestress.cfm.
25. Ibid.
26. See Sternberg, 1992, note 18.
27. Harvard Mental Health Letter, 'The mind and the immune system' (April 2002), available at: http://www.health.harvard.edu/medline/mental/M0402a.html.
28. D. Hamerman, 'Toward an understanding of frailty', *Annals of Internal Medicine*, 130, 1999, 945–50.
29. L. Ferrucci et al., 'Serum IL-6 level and the development of disability in older persons', *J. Am. Geriatr. Soc.*, 47, 1999, 639–46.
30. Ibid.
31. Ibid.
32. W. B. Ershler and E. T. Keller, 'Age-associated increased interleukin-6 gene expression, late-life diseases, and frailty', *Ann. Rev. Med.*, 51, 2000, 245–70.
33. Ibid.
34. Ibid.
35. R. Kalra et al., 'Subclinical doses of the nerve gas sarin impair T cell responses through the autonomic nervous system', *Toxicology and Applied Pharmacology*, 184, 2002, 82–7.
36. Ibid.
37. E. M. Sternberg, *SNIB: Section on Neuroendocrine Immunology and Behavior* (2003), available at: http://intramural.nimh.nih.gov/snib/.
38. M. E. Poynter and R. A. Daynes, 'PPRAα activation: A drink from the fountain of youth?', *Biologist*, 51 (1), 2004, 27–31.
39. E. M. Sternberg, 'Neuroendocrine regulation of autoimmune/inflammatory disease', *Journal of Endocrinology*, 169, 2001, 429–35.
40. M. J. Coghlan et al., 'Selective glucocorticoid receptor modulators', *Annual Reports in Medicinal Chemistry*, 37, 2002, 167–75.
41. T. L. Bale and W. W. Vale, 'CRF and CRF receptors: Role in stress responsivity and other behaviours', *Ann. Rev. Pharmacol. Toxicol*, 44, 2004, 525–57.
42. Ibid.

43. N. Kawashima and A.W. Kusnecov, 'Effects of staphylococcal enterotoxin A on pituitary-adrenal activation and neophobic behaviour in the C57 BL/6 mouse', *Journal of Neuroimmunology*, 123, 2002, 41–9.
44. T. Kaneta and A.W. Kusnecov, 'The role of central corticotropin-releasing hormone in the anorexic and endocrine effects of the bacterial T cell superantigen, *Staphylococcal enterotoxin A', Brain, Behaviour, and Immunity*, 19, 2005, 138–46.
45. A. Rossi-George et al., 'Neuronal, endocrine, and anorexic responses to the T-cell superantigen Staphylococcal enterotoxin A: Dependence on tumor necrosis factor-α', *Journal of Neuroscience*, 25, 2005, 5314–22.
46. L. Steinman, 'Elaborate interactions between the immune and nervous systems', *Nature Immunology*, 5, 2004, 575–81.
47. A. Inui, 'Cytokines and sickness behaviour: implications from knockout animal models', *Trends in Immunology*, 22, 2001, 469–73.
48. M. A. Petty and E. H. Lo, 'Junctional complexes of the blood-brain barrier: permeability changes in neuroinflammation', *Progress in Neurobiology*, 68, 2002, 311–23.
49. J. Licinio and P. Frost, 'The neuroimmune-endocrine axis: pathophysiological implications for the central nervous system cytokines and hypothalamus-pituitary-adrenal hormone dynamics', *Brazilian Journal of Medical and Biological Research*, 33, 2000, 1141–8; Steinman (2004), note 46.
50. M. Dardenne and W. Savino, 'Control of thymus physiology by peptidic hormones and neuropeptides', *Immunology Today*, 15, 1994, 518–26; Inui (2001), note 47.
51. S.-M. Tsai et al., 'Pyrogens enhance β-endorphin release in hypothalamus and trigger fever that can be attenuated by buprenorphine', *Journal of Pharmacological Science*, 93, 2003, 155–62.
52. Licinio and Frost (2000), note 49; Steinman (2004), note 46.
53. L. M. Boulanger and C. J. Shatz, 'Immune signalling in neural development, synaptic plasticity and disease', *Nature Reviews Neuroscience*, 5, 2004, 521–31.
54. J. E. Blalock, 'The syntax of immune-neuroendocrine communication', *Immunology Today*, 15, 1994, 504–11.
55. Licinio and Frost (2000), note 49.
56. H. Wekerle, A. Flügel and H. Neumann, 'Neuronal control of the immune response in the central nervous system: from pathogenesis to therapy', in P. Patterson, C. Kordon and Y. Christen (eds), *Neuro-Immune Interactions in Neurologic and Psychiatric Disorders* (Berlin: Springer, 2000), pp.111–23; M. Kubera, and M. Maes, 'Serotonin-immune interactions in major depression', in Patterson, Kordon and Christen (eds), *Neuro-Immune Interactions*, pp.79–87.
57. S. Y. Felten and D. L. Felten, 'Neural-immune interactions', *Progress in Brain Research*, 100, 1994, 157–62; R. H. Straub et al., 'Dialogue between the CNS and the immune system in lymphoid organs', *Immunology Today*, 19, 1998, 409–13.
58. Steinman (2004), note 46; Straub et al. (1998), note 57.

59. Steinman (2004), note 46.
60. M. B. Tanzola et al., 'Mast cells exert effects outside the central nervous system to influence experimental allergic encephalomyelitis disease course', *Journal of Immunology*, 171, 2003, 4385–91.
61. C. Lock, et al., 'Gene microarray analysis of multiple sclerosis lesions yields new targets validated in autoimmune encephalomyelitis', *Nature Medicine*, 8, 2002, 500–08.
62. Inui (2001), note 47.
63. A. K. Abbas, A. H. Lichtman and J. S. Pober, *Cellular and Molecular Immunology* (Philadelphia: W.B. Saunders Company, 1997).
64. USAMRIID, 'Basic studies seeking generic medical countermeasures against agents of biological origin', *Annual Report for Fiscal Year 1987*, p.19, reported in B. Rosenberg and G. Burck, 'Verification of compliance with the Biological Weapons Convention', in S. Wright (ed.), *Preventing a Biological Arms Race* (Cambridge, MA: MIT Press, 1990), pp.301–29.
65. Straub et al. (1998), note 57.
66. Licinio and Frost (2000), note 49.
67. See USAMRIID (1987), note 64.
68. R. A. Goldsby et.al. *Immunology*, 5th edition (New York: W. H. Freeman and Company, 2003).
69. K. Alibeck and S. Handelman, *Biohazard. The Chilling True Story of the Largest Biological Weapons Program in the World – Told from Inside by the Man Who Ran It* (New York: Random House Inc., 1999).
70. C. Lundberg, S. J. Jungles and R. C. Mulligan, 'Direct delivery of leptin to the hypothalamus using recombinant adeno-associated virus vectors results in increased therapeutic efficacy', *Nature Biotechnology*, 19, 2001, 169–72.
71. Inui (2001), note 47.
72. See Boulanger and Shatz, 2004, note 53.
73. P. Esposito et al., 'Acute stress increases permeability of the blood-brain barrier through activation of brain mast cells', *Brain Research*, 888, 2001, 117–27.

7 Assessing the Adequacy of the CBW Prohibition Regimes for the Challenges of the 21st Century

1. E. A. Fenn, *Pox Americana. The Great Smallpox Epidemic of 1775–82* (New York: Hill and Wang, 2001).
2. M. R. Dando, V. Nathanson and M. Darvell, *Biotechnology, Weapons and Humanity* (London: Harwood Academic Publishers, 1999).
3. W. S. Carus, *Bioterrorism and Biocrimes. The Illicit Use of Biological Agents Since 1900*, Center for Counterproliferation Research Working Paper (Washington, DC: National Defense University, February 2001).
4. E. Geissler and J. E. v. C. Moon (eds), *Biological and Toxin Weapons: Research, Development and Use from the Middle Ages to 1945*, SIPRI Chemical & Biological Warfare Studies No.18 (Oxford: Oxford University Press, 1999).

5. R. J. Jackson et al., 'Expression of mouse interleukin-4 by recombinant ectromelia virus suppresses cytolytic lymphocyte responses and overcomes genetic resistance to mousepox', *J. Virol.*, 75, 2001, 1205–10.
6. J. Cello, et al., 'Chemical synthesis of poliovirus cDNA: Generation of infectious virus in the absence of natural template', *Science*, 297, 2002, 1016–18.
7. A. M. Rosengard et al., 'Variola virus immune evasion design: Expression of a highly efficient inhibitor of human complement', *Proc. Natl. Acad. Sci.*, 99, 2002, 8808–13.
8. M. R. Dando, *Biotechology, Weapons and Humanity II* (London: British Medical Association, 2004).
9. G. Poste, 'Advances in biotechnology: Promise or peril (2002), available at: www.hopkins-defense.org/sympcast/transcripts.trans_poste.html.
10. M. Meselson, 'The problem of biological weapons', Presentation given at the 1818th Stated Meeting of the American Academy of Arts and Sciences (Cambridge, MA: 13 January 1999).
11. G. L. Epstein, *Global Evolution of Dual-Use Biotechnology*, CSIS Report (Washington, DC: CSIS, April 2005).
12. J. Tian et al., 'Accurate multiplex gene synthesis from programmable DNA microchips', *Nature*, 432, 2004, 1050–4.
13. P. Carr et al., 'Protein-mediated error correction for de novo DNA synthesis', *Nucleic Acids Research*, 32 (20), 2004, e 162.
14. G. Church, *A Synthetic Biohazard Non-Proliferation Project* (2004), available at: http://arep.med.harvard.edu.
15. K. Pekrun et al., 'Evolution of a human immunodeficiency virus type 1 variant with enhanced replication in Pig-tailed Macaque cells by DNA shuffling', *J. Virol.*, 76, 2002, 2924–35.
16. J. Lieberman et al., 'Interfering with disease; Opportunities and roadblocks to harnessing RNA interference', *Trends in Molecular Medicine*, 9, 2003, 397–403.
17. R. M. Krug, 'The potential use of influenza virus as an agent for bioterrorism', *Antiviral Research*, 57, 2003, 147–50.
18. Y. Yokobayashi et al., 'Directed evolution of a genetic circuit', *Proc. Natl. Acad. Sci.*, 99, 2002, 16587–91.
19. J. Garcia-Ojalvo et al., 'Modelling a synthetic multicellular clock: Repressilators coupled by quorum sensing', *Proc. Natl. Acad. Sci.*, 101, 2004, 10955–60.
20. M. Allert et al., 'Computational design of receptors for an organophosphate surrogate of the nerve agent soman', *Proc. Natl. Acad. Sci.*, 101, 2004, 7907–12.
21. A. Goldbeter, 'Computational biology: A propagating wave of interest', *Curr. Biol.*, 14, 2004, R601–02.
22. G. Apic et al., 'Illuminating drug discovery with biological pathways', *FEBS Letters*, 579, 2005, 1872–7.
23. G. Poste, op. cit. reference 12.
24. P. Aloy and R. B. Russell, 'Structure-based systems biology: A zoom lens for the cell', *FEBS Letters*, 579, 2005, 1854–8.

25. P. Aloy and R. B. Russell, 'Ten thousand interactions for the molecular biologist', *Nature Biotechnology*, 22, 2004, 1317–21.
26. See Chapter 3, section 3.2.1.
27. Depositary States, *New Scientific and Technological Developments Relevant to The Convention on the Prohibition of the Development, Production and Stockpiling of Bacteriological (Biological) and Toxin Weapons and on Their Destruction.* Document BWC/CONF.I/5 (Geneva: United Nations, 6 February 1980).
28. *New Scientific and Technological Developments Relevant to The Convention on the Prohibition of the Development, Production and Stockpiling of Bacteriological (Biological) and Toxin Weapons and on Their Destruction*, Document BWC/CONF.III/4 (Geneva: United Nations, 26 August 1991).
29. Canada, *Novel Toxins and Bioregulators: the Emerging Scientific and Technological Issues Relating to Verification and the Biological and Toxin Weapons Convention* (Ottawa, September 1991).
30. *New Scientific and Technological Developments Relevant to The Convention on the Prohibition of the Development, Production and Stockpiling of Bacteriological (Biological) and Toxin Weapons and on Their Destruction*, Document BWC/CONF.V/4 (Geneva: United Nations, 14 September 2001).
31. *New Scientific and Technological Developments Relevant to The Convention on the Prohibition of the Development, Production and Stockpiling of Bacteriological (Biological) and Toxin Weapons and on Their Destruction*, Document BWC/CONF.V/4 Add.1, Geneva: United Nations, 26 October 2001).
32. W. Cohen, *Proliferation: Threat and Response* (Washington, DC: Department of Defense, 1997).
33. A. Kelle, *Bioterrorism and the Securitization of Public Health in the United States of America – Implications for Public Health and Biological Weapons Arms Control*, Bradford Regime Review Paper No.2 (Bradford: University of Bradford, July 2005), p.2, available at: http://www.brad.ac.uk/acad/sbtwc/regrev/Kelle_SecuritizationinUS.pdf.
34. *Note by the Director General. Report of the Scientific Advisory Board on Developments in Science and Technology*, OPCW document RC-1/DG.2, The Hague, 23 April 2003, p.15.

8 Conclusion: Towards an Overarching Framework for Biochemical Controls

1. G. S. Pearson, 'Prospects for chemical and biological arms control: the web of deterrence', *The Washington Quarterly*, 16 (2), 1993, 145–62.
2. M. R. Dando, V. Nathanson and M. Darvell, *Biotechnology, Weapons and Humanity* (London: Harwood Academic, 1999).
3. S. Spence, *Achieving Effective Action on Universality and National Implementation: the CWC Experience*, Briefing Paper No.13 (Bradford: University of Bradford, April 2005).
4. United Nations, *Meeting of the States Parties to the Convention on the Prohibition of the Development, Production and Stockpiling of Bacteriological*

(Biological) and Toxin Weapons and on their Destruction: Report of the Meeting of States Parties, Document BWC/MSP/2003/4 (Vol.I) (Geneva: United Nations, 26 November 2003).

5. Ibid., reference 4.
6. Ibid., reference 4.
7. Ibid., reference 4.
8. G. S. Pearson, 'Security and oversight of pathogenic microorganisms and toxins', *Chemical and Biological Weapons Conventions Bulletin*, 60, 2003, 6–15.
9. United Nations, *Meeting of the States Parties to the Convention on the Prohibition of the Development, Production and Stockpiling of Bacteriological (Biological) and Toxin Weapons and on their Destruction: Report of the Meeting of States Parties*, Document BWC/MSP/2004/3 (Geneva: 14 December 2004).
10. Ibid., reference 9.
11. United Nations, *Meeting of the States Parties to the Convention on the Prohibition of the Development, Production and Stockpiling of Bacteriological (Biological) and Toxin Weapons and on their Destruction: Report of the Meeting of States Parties*, Document BWC/MSP/2005/MX/3 (Geneva: United Nations, 5 August 2005).
12. G. S. Pearson, 'Report from Geneva No. 23: The Biological Weapons Convention Meeting of Experts', *The Chemical and Biological Weapons Conventions Bulletin*, 68, 2005, 12–19.
13. Ibid., reference 12.
14. M. R. Dando and B. Rappert, *Codes of Conduct for the Life Sciences: Some Insights from UK Academia*, Briefing Paper No.16 (Bradford: University of Bradford, May 2005).
15. Committee on Research Standards and Practices to Prevent the Destructive Application of Biology, *Biotechnology Research in an Age of Terrorism: Confronting the Dual-Use Dilemma* (Washington, DC: The National Academies Press, 2003).
16. J. D. Steinbrunner and E. D. Harris, 'Controlling dangerous pathogens', *Issues in Science and Technology*, 19 (3), 2003, available at: http://www.issues.org/issues/19.3/steinbruner.htm.
17. M. R. Dando and M. L. Wheelis, 'Back to bioweapons', *Bulletin of the Atomic Scientists*, 59 (1), 2003, 40–6.
18. E. D. Harris and J. D. Steinbrunner, 'Scientific openness and national security', *Chemical and Biological Conventions Bulletin*, 67, 2005, 1–6.
19. Royal Society, *Royal Society Submission to the Foreign and Commonwealth Office Green Paper on Strengthening the Biological and Toxin Weapons Convention*, Policy Document 25/02 (London: Royal Society, 2002).
20. K. D. Ward, 'The BWC Protocol: Mandate for Failure', *The Nonproliferation Review*, 11 (2), 2004, 183–99.
21. US Department of State, Bureau of Public Affairs, *Proliferation Security Initiative*, Washington, DC., 15 September 2003, available at: http://www.state.gov/documents/organization/24252.pdf.
22. Ibid.

23. T. D. Lehrman, 'Rethinking interdiction: the future of the Proliferation Security Initiative', *The Nonproliferation Review*, 11 (2), 2004, 1–45.
24. Ibid, p.27.
25. United Nations, Security Council, *Resolution 1540 (2004). Adopted by the Security Council at its 4956th Meeting, on 28 April 2004*, available at: http://disarmament2.un.org/Committee1540/Res1540(E).pdf.
26. Idem.
27. A more detailed account of the negotiation, content and implications of UN Security Council Resolution 1540 can be found in M. Datan, 'Security Council Resolution 1540: WMD and non-state traficking', *Disarmament Diplomacy*, 79, 2005; C. Craft, *Challenges of UNSCR 1540: Questions about International Export Controls*, CITS Briefs (Athens: University of Georgia, 2004).
28. J. B. Tucker, *Biosecurity: Limiting Terrorist Access to Deadly Pathogens*, Peaceworks No.52 (Washington, DC: United States Institute of Peace, November 2003).
29. Ibid, p.5.
30. Ibid, pp.29–34.
31. See: http://www.sussex.ac.uk/Units/spru/hsp/CRIMpreambleFeb04.htm.
32. Ibid.
33. See Spence, note 3.
34. See: http://www.sussex.ac.uk/Units/spru/hsp/CRIMpreambleFeb04.htm.
35. N. A. Sims, *Remedies for the Institutional Deficit of the BTWC: Proposals for the Sixth Review Conference*, Review Conference Paper No.12 (Bradford: University of Bradford, March 2005).
36 Ibid, p.4.
37. See: http://unfccc.int/2860.php.
38. See: http://www.who.int/tobacco/framework/en/.
39. M. L. Wheelis, ' "Nonlethal" chemical weapons – a Faustian bargain', *Issues in Science and Technology*, Spring 2003, 74–8, available at: http://www.issues.org/issues/19.3/wheelis.htm.

Index